Joseph Henry:
HIS LIFE AND WORK

Joseph Henry, from a photograph taken about 1875.

Joseph Henry
HIS LIFE AND WORK
By Thomas Coulson

PRINCETON
PRINCETON UNIVERSITY PRESS
1950

Copyright, 1950, by Princeton University Press
London: Geoffrey Cumberlege, Oxford University Press

Printed in the United States of America
By Vail-Ballou Press, Inc., Binghamton, N. Y.

TO RICHARD WINGATE LLOYD

TO RICHARD WINSLEY LLOYD

PREFACE

THIS book is an attempt to describe the life and work of Joseph Henry, written in the hope that it may assist in rescuing him from the neglect into which he has fallen and in restoring him to a rightful place among the gallery of America's great men. The author is conscious of his inadequacy for the task he has undertaken and makes ready acknowledgement of the help he has received in the accomplishment of the work. Special acknowledgement is due to Mr. Herbert S. Bailey, Jr., of Princeton, who gave invaluable advice, encouragement, and more tangible aid in permitting the author to make use of his notes on Henry's experiments with the electric magnet. To him the author expresses his gratitude.

Mr. Walter Pertuch, librarian of The Franklin Institute, proved to be a colleague of unfailing resource and patience. His aid in some research problems must also be gratefully acknowledged. Mr. Britton Holmes of the Albany Academy was also most helpful.

A bibliography is supplied, but reference must be made here to the manuscript resources of the Smithsonian Institution, Henry's great monument. Most of the quotations from letters and all references to the European journey of 1837 (when other sources are not given) were derived from this source. The labor of examining this material was greatly simplified through the courtesy of Mr. Jack Hollingsworth, of Drexel Hill, Pennsylvania, who most generously placed at the author's disposal his microfilm reproduction of this material. Grateful acknowledgement is made to Mr. Hollingsworth and to the Smithsonian Institution for permitting the photographing of this material. The author accepts sole responsibility for any inaccuracy which may have crept into the manuscript.

July 1949 T. COULSON
The Franklin Institute
Philadelphia

CONTENTS

PREFACE		vii
I.	EARLY DAYS	3
II.	WORK AT ALBANY	25
III.	THE DISCOVERY OF INDUCTION	65
IV.	PRINCETON	96
V.	FOUNDING THE SMITHSONIAN INSTITUTE	169
VI.	THE TELEGRAPH CONTROVERSY	208
VII.	THE RIPENING YEARS	235
VIII.	ORGANIZER OF SCIENCE	270
IX.	JOSEPH HENRY, THE MAN	284
X.	CONSUMMATION	318
XI.	HENRY'S PLACE IN SCIENCE	330
APPENDIX		341
BIBLIOGRAPHY		344
INDEX		347

Joseph Henry:
HIS LIFE AND WORK

EARLY DAYS

Joseph Henry is a strangely neglected figure, and yet the story of his life and work has many of those elements which endear a man to the American mind. His was a true rise from poverty to, not riches, but fame. The son of a day laborer in Albany, he was aimlessly drifting until chance guided him to a career. At a time when America had few professional scientists he engaged in scientific investigations which bore the most fruitful results. The difficulties he had to overcome were impressive. He worked in a frontier country remote from the current of European scientific thought and with only occasional information regarding the labors of others. In those early days his time was fully occupied with teaching and his experimental work was largely limited to the summer vacations. He labored under severe financial restrictions. His results were achieved by great industry in the application of his genius for practical experimentation, his inventive insight, his power for exact observation, and a sense of perspective and proportion.

In spite of the handicaps under which he worked, Henry's achievements were sufficiently great to distinguish him as America's foremost physical scientist. The brief periods in which he was permitted to perform original experiments were marked from the first with important discoveries. He converted the electromagnet from a toy into an instrument with vast potentialities. With the new power he had created he moved from one discovery to another and, through an idiosyncrasy, just failed to announce the discovery of electromagnetic induction. But for this omission, his name would have been as familiar to Americans as that of Faraday to Englishmen or of Pasteur to

Frenchmen. His discovery of electromagnetic self-induction is a scientific achievement of the first importance; and his name has been immortalized as the unit of electromagnetic induction, the henry.

Even without this distinction his contributions to knowledge were substantial enough to earn for him a permanent place in history. With his method of insulating wire it could be made into the large coils necessary for powerful electromagnets. Henry used his magnets to investigate the principles of electromagnetism from which sprang the dynamo and the motor. He saw the possibilities more clearly than anyone else and gave a direction to their principal applications when he devised a reciprocating machine which operated by electromagnetism and contrived an apparatus which would ring a bell at the end of a mile of wire that was the forerunner of the electric telegraph.

These achievements brought him recognition and led to his translation to a post of greater distinction at Princeton. Here, he continued his investigations under more favorable conditions and attained brilliant results. His further discoveries produced invaluable electric appliances such as the relay and the transformer. By his research into the variation of induction between coils without separation, and in the action of induction at a distance he prepared the way for radio telegraphy and telephony. Between these major discoveries he made others of only slightly less importance to the development of electrical science, and still found time to make investigations which advanced knowledge in other fields.

At a time when he appeared to be entering the most productive years of his maturity Henry renounced active research work in order to undertake a new task when he accepted the appointment to organize the Smithsonian Institution. Henry had begun his career under adverse conditions and at a time when scientific effort was without coordination. Remembering the obstacles which stood in his way as a student, he perceived that in this new post he could exercise an influence which would stimulate the scientific effort of the nation. From 1846 to the time of his death in 1878 he devoted his energies to the removal of those obstacles through the promotion of societies, foundations, and publications which would eliminate much of the accidentalism from the work of scientists and would encourage the collective effort necessary to promote the increase and diffusion of knowledge, avoid duplication of effort, and obtain recognition for the man of science. He made the Smithsonian Institution the greatest influence upon American education for almost a century.

No man did more than Joseph Henry did to elevate the position of the scientific worker. When he died in his eighty-sixth year his funeral was one of the most impressive ever seen in Washington. Yet the loss of this great man was felt most by the relatively few scientists, legislators, and educators who understood his work. The man in the street was barely sensible of his passing because Henry had no interest in publicity or fortune, the two greatest criteria of popularity. His failure to capitalize upon his discoveries and his omission to profit from the electrical utilities he had helped to make possible presents an interesting study in human behavior. Perhaps it is the difficulty of explaining him that has left him such a vague figure in American history.

On June 16, 1775, the day of the battle of Bunker Hill, a vessel arrived in New York harbor bearing a new group of settlers from Scotland and England. The names of the passengers have been forgotten, save for two families, the Hendries and the Alexanders, whose identity was to be preserved through the rise to fame of a yet unborn child. An inarticulate group, they recorded scarcely a word of their origin, except that the Hendries were of an Argylshire family distantly related to the Earls of Stirling.

New York, then a little town of thirty thousand inhabitants, with its brick buildings and tiled roofs, its twenty churches and dancing academies, the luxurious ornamentation on the more fashionable houses, its progressive and confident citizens, might have been supposed to attract the newcomers from austere Scotland. The immigrants may have observed the outward signs of prosperity and refinement without perceiving that the minds of the citizens were unquiet over political affairs.

However, undeterred by the attractions of the city and by any disturbing news of the conflict in the North, the new arrivals turned their backs upon the "metropolis and grand mart" as the inhabitants proudly called it, and moved upstate, where lay the patroonships granted by the Dutch West India Company. Here were settled a few fellow Scots who had fled their native land after the battle of Killiecrankie had filled it with despair and woe. Arrived at Albany the two families separated.

William Hendrie and his family settled on a farm in Delaware County, where they became submerged in the mingling of five nationalities. William, as was not uncommon among Scots, was well read in Presbyterian theology and Scottish history, but had given no great attention to other fields of learning. He was loyal to the old Scottish monarchy, a loyalty which did not diminish with the passage of the years in a country which was not distinguished for its respect for kings. He lived to be nearly ninety and, in later life, found delight in telling how he had seen Charles Stuart ride into Glasgow in 1745.

William Hendrie, conservative though he was, did not oppose all change. It cannot have been long after his arrival in this country that he approved of a change in his name. The name Hendrie distinguishes its bearer instantly as of Scottish ancestry, and for this reason one would have thought that William would have wished to retain it as a link with the land where he had been born. However, he was not long in consenting to a modification of the name to Henry, probably from deference to his neighbors' inability to pronounce the uncommon variant of a familiar name.

Joseph Henry did not approve of this change. He wrote to an admirer who had called his son after the scientist: "I did not object to Henry as a first name; although I have been sorry that my grandfather, in coming from Scotland, substituted it for Hendrie, a much less common and therefore more distinctive name."

Joseph Henry's maternal relatives, the Alexanders, had a more exciting existence. Hugh Alexander set himself up as a miller after constructing his own mill, even to the grinding stones, with the aid of his sons. During the course of the Revolutionary War, the family was driven from its home by Indians, but Hugh became an artificer in the continental army. When a strained peace descended upon the Hudson Valley after the surrender of Burgoyne's army, Hugh Alexander and his family resumed their normal lives, although it would seem that the mill had been destroyed, for Hugh appears then as a manufacturer of salt at Salina.

When peace finally came to the country, the relations between the two immigrant families were restored. Romance which had sprung up between two of the children during the voyage from Scotland grew into full flower, and William Henry the younger married Ann Alexander. The newly married couple set up housekeeping in a small clapboard home on South Pearl Street in Albany. The younger

William Henry was not endowed with sound health. He was soon to die, and can have been little better than a vague memory to his children.

Ann, his wife, left a clearer mark. Those who recalled her appearance described her as *petite,* fair, and possessed of pure Grecian features. Since the early death of her husband enabled her influence to be more deeply engraved upon her children's character, it is of some importance to know what kind of woman was Ann Alexander Henry.

Everyone who speaks of her mentions her strong character. Even in the poorest circumstances she was reputed to have retained the demeanor of a gentlewoman, which is not an uncommon trait among Scottish women of humble birth. Ann Henry was also respected for the refinement of her mind. Her rigid puritanism had a deep influence on young Joseph. His rejection of wealth in order to embrace a career which offered few rewards, and the modesty which caused him to give praise and credit to others while claiming the minimum for himself, are characteristics supposed to have been the result of his mother's stern training.

The Scottish life was compounded of a niggard soil, mists, granite, and Calvinism. There was something hard and cold, recalling the granite, in the national character, but there were also reserves of warmth, feeling, and perception that were equally strong in Henry's character.

There is some doubt about the date of Joseph Henry's birth. The baptismal register of the First Presbyterian Church in Albany records the date as December 9, 1797. But when his cousin and confidant Stephen Alexander was consulted, he was emphatic that this register entry was an error. In a letter to Asa Gray [1] he said: "I had learned years ago, the date (as to the year) of Professor Henry's birth, and compared his age with my own. The difference was what it was not likely I would forget. But to set all convincingly right, I wrote to the friend who had charge of the Professor's life insurance; and his investigation brought out the same result with that which I had insisted on. Professor Henry was born in 1799. That is certain. The statements, in addition, say in Dec. of that year."

Thus, we are thrown into confusion at the very outset of his career.

The earliest years of Joseph Henry's life were spent in modest, if not straitened circumstances. For some reason which is not easy to

---
[1] Preserved at the Gray Herbarium, Harvard University.

understand, Mary Henry, Joseph's daughter who appears to have contemplated writing a biography of her father, described her grandfather as a seafaring man. The description is open to doubt, for the contemporary records say that William Henry was a day laborer, and this at a time when most men in the country were wresting a living from small freeholds and calling themselves farmers.

When Joseph was nearly seven years old the domestic situation threatened to slip beyond the capable Ann Henry's control. Apparently, in order to obtain a partial relief from family cares and to gain time to nurse her ailing husband Ann was compelled to send the boy to Galway, some thirty-six miles away, where he was to live with her stepmother and her twin brother, John.

Here the boy attended the village school, taught by Israel Phelps. Phelps cannot have been a discriminating pedagogue, eager to seek out and encourage an expanding mind. He made no observable impression on this lump of plastic material. Nor was there anything of the infant prodigy about the young Joseph. Whatever dormant powers he possessed needed awakening, and Phelps was not the man to do this. Joseph attended school because it was inescapable. He displayed no desire to acquire knowledge. If he learned the rudiments of reading, writing, and arithmetic, he certainly did not acquire the sense that learning was the object that made the scholar a leader among men.

Having reached the age of ten, and being possessed of only the most elementary education, he was taken into employment at the general store, run by a man called Broderick. The storekeeper developed a kindly feeling toward his employee. He had a respect for education, even of the kind dispensed by Israel Phelps. He saw to it that Joseph attended afternoon classes after spending his morning hours in the store.

Joseph Henry is described at this period as being remarkable for his good looks, delicate complexion, slight figure, and vivacious nature. His lively temperament made him a general favorite. He was popular with the other boys of the village, and was usually to be found in the center of a group. He must have mingled freely, too, with his elders, for the village store was the center of social activity.

Galway was a small place; life there was neither gay nor exciting. Such an environment can be a severe test of a growing lad's character. It can cause him to vegetate, or it can cause him to rebel and break the bonds of his surroundings. Given imagination, a boy is almost sure to break away. Young Joseph had plenty of imagination. His individuality began to manifest itself in a way that the villagers did not understand. He was occasionally accused of day-dreaming, which was set down to idleness.

The primitive educational system of the day failed to take into account the training of the imagination, so that the future scientist had to cultivate it in his own way. The good people among whom he lived did not realize that imitation and imagination develop more quickly than the reasoning faculty. Tedious family chores and the not too onerous duties at the store having been disposed of, young Joseph would sometimes avoid his companions and roam off alone into the fields. An impatient grandmother might call fruitlessly for him while he lay, half-buried in the grass, as he afterwards recalled, speculating upon the creation of the universe.

The dreamy spirit which marked his youth would have indicated the development of an artist, poet, or actor. The next phase of his life served to confirm this view, but one should not seek too deeply in those boyhood days for evidences of later sagacity and character. Boys are characteristically fitful and unfinished, yet one singular streak in Henry's character was apparent all through his life. In later years, when he was the executive officer of a national institution, he was often accused of straddling, of being unable to arrive at a decision. He was aware of this tendency. Once, when reproached with failing to meet an issue squarely and promptly, he recalled an incident of those Galway days which showed that he had been plagued with indecision all his life.

This incident occurred when he was permitted to exercise an independent choice of new shoes. He was not tantalized by any great variety of styles. The selection lay simply between shoes with round or with square toes. Unable to make up his mind at the moment, Joseph advised the village shoemaker to proceed with the preliminaries while he considered the problem of the toes.

Daily he visited the craftsman to observe progress, but left after each inspection with judgement reserved upon the toes. No decision had been reached when Joseph went one day for his customary conference, only to be presented with the most remarkable pair of shoes he was ever to own. One had a square toe and the other a round one!

He had an unconventional introduction to the joys of reading.

Among his few personal possessions was a pet rabbit. One day the animal escaped from its inclosure. In pursuit the boy was compelled to crawl under the village church. Suddenly his attention was distracted by the appearance of chinks of light overhead. He decided to investigate and found some boards in the church flooring that were loose enough to yield to pressure. He was able to make a hole through which he wriggled into the building. To his surprise, he found himself in a small room he did not know existed. He had stumbled upon the village library.

The boy had displayed no interest in the few books that came his way in school, but this was different. For the first time in his life he was confronted with books he was under no compulsion to read, and he began to sample the shelves.

His first choice of reading matter would not strike a youth of today as a promising introduction to the riches of literature. The volume he selected was Sir Philip Brook's *The Fool of Quality,* one of those moralistic romances which awards riches and happiness only to the virtuous. The book had gained the approval of John Wesley, a stern rather than a discriminating critic.

The probability is that Henry was less attracted by the moral issues raised in the book than to the passing likeness of himself in the hero. The fortunes of two sons of an English peer were followed. The elder son was trained to the succession of his father's high place in society and developed lazy and dissolute habits. The younger son, the hero, was placed as a foster child with a farmer's wife, where he acquired only sturdy qualities. The book was filled with disquisitions on sociology, economics, and religion of the elementary kind which Joseph must have heard expressed around Broderick's stove. Through his hero the author spoke against the oppression of the poor working class, pointed out that poverty arose from ignorance rather than from idleness or incapacity, and advocated social progress through the digging of canals and improved agricultural methods.

Having read as long as he dared, the boy replaced the volume and left the room by the way he had entered. He returned on successive days until he had learned the outcome of the story. Although the work was hardly of the nature to capture the interest of a lively immature mind, it made a deep impression upon this reader. Expressions made familiar by repetition in the village store gave the book an air of reality. With certain other resemblances to his own life, his uncritical young mind would readily assume the remainder of the story to be a close approximation to life.

The chance reading of *The Fool of Quality* ought not to be regarded as an isolated incident in an otherwise dull chronicle of childhood. It was one of the few moulding influences of Henry's life. He had been admitted to a new world, the realm of books, and it never ceased to interest him.

The discovery of this book stimulated the desire to read more. He returned to the library by his secret entrance on several occasions. Fortunately for him, the church library was more than an arsenal of piety. He found other novels on its shelves, and these he eagerly devoured. Daily visits to the library by way of its structural foundations could not fail to attract attention, but the first person to learn of his unconventional behavior was, luckily, Mr. Broderick, his employer, himself a great reader. Broderick arranged for Henry's admission to the library in more orthodox fashion, and the pair had the additional enjoyment of discussing the books they read.

His interest in books enhanced Joseph's standing with his youthful companions who did not enjoy the privilege of using the library. Now, instead of being found flat on his back questioning the clouds he exercised his memory and imagination by beguiling other boys with the stories he was reading. Then, on visiting Albany to see his sick father, he had been taken to see a play. He delighted his Galway companions on his return by reenacting the entire play, impartially playing the parts of all the actors.

Meanwhile, the boy had grown. His father had died. Death, too, had removed his Uncle John with whom he had been living. Family readjustments had to be made, and, as Joseph was approaching the age of fourteen, it was time for him to be thinking of earning a living. He returned to Albany.

In the first quarter of the nineteenth century, Albany was a pleasant little town where the majority of the houses had cross-stepped gables and most of the inhabitants were of Dutch descent. The place had shared in the prosperity of all river towns, since the rivers were the principal highways of commerce. It was the active seat of the State administration, and it was surrounded by large estates, whose owners exercised a strong influence upon the State government. The political atmosphere was reminiscent of feudalism.

Soon after his return to Albany, Joseph was apprenticed to John F. Doty, a watchmaker and silversmith. One cannot say that he developed a keen interest in the fine work with file and crucible, but what he learned was to serve him well in making apparatus at a later stage. Very probably he disliked the occupation because of the confinement, and found its genteel dignity a poor substitute for the virile flow of life around the village store, with its tributary streams of idlers and gossips.

Moreover, it was not a prosperous time for a luxury trade. The nation was suffering from the economic effects of the Embargo Acts. The country was headed for the financial panic of 1819 which left the West prostrate and the nation staggering. Doty battled against the adverse economic tide only long enough to give his apprentice two years of training before giving up his business. In parting, he expressed the opinion that Joseph was too dull to make a silversmith.

This dismal judgement was undoubtedly based upon the boy's lack of interest in his tasks. The employer probably knew that the boy's mind was inflamed by romantic ideas gathered from his reading, and from the influence of what was probably at first a clandestine interest in the theatre.

In the years 1813–1816 Albany possessed an unusually good theatre under the management of John Bernard, an English comedian of more than average ability and author of several books on the early American stage. He had gathered about him a stock company that included several actors who became prominent in their profession, such as Samuel Drake and his wife, Ann Denney, Henry Placide, and Norah Ludlow.

How Joseph Henry came to make these excursions from an austere home into the pagan world of the theatre is one of the inexplicable things of his life. It is hardly consistent with the training instilled by his widowed mother, if she was as rigidly pious as she is pictured. Spirit may have salted their poor fare, and may have helped to make up for short commons, but it is difficult to reconcile Henry's frequenting of the theatre with the somber Calvinism of his home. It was not so long before that Jonathan Edwards had preached damnation for the giddy. Perhaps Ann Henry's religious creed was not as inflexible as we have been led to believe.

More difficult to understand is where the boy found the money to witness many plays. The prices of admission to the theatre were not so low as to permit frequent attendance by a mere apprentice and the son of a widowed mother who had been reduced to renting rooms to boarders. The explanation to this point may be found in his presence behind the scenes, a privilege for which he did not have to pay if he made himself useful.

This experience fired Henry with a desire to make the stage his career. For more than a year the theatre absorbed most of his leisure. He joined, if he did not actually organize, a society of amateurs calling themselves "The Rostrum." Since he followed no regular occupation, he devoted himself to acting, writing and producing plays for this group. He is said to have dramatized a favorite novel and to have written an original comedy, both of which were acted by the players. At the age of sixteen, he was elected president of "The Rostrum." He was in a fair way to make a place for himself among the "periwig-pated fools who tear a passion to tatters."

This episode of his life is not to be accounted altogether a waste of time, effort, and talent, for it enabled him to cultivate an ease of platform speaking and that effective form of address which was later to distinguish him as a lecturer.

At this critical point in young Henry's life, we may see him more clearly against the pattern of events at the beginning of the nineteenth century. By the year 1810 a steady procession of wagons, horses, and men had begun to move along the turnpikes leading to the West. Boys who lived on farms where a new crop of stones came up every spring were lured by the deeper richer soil of Ohio. Boys who were not attached to farms may not have been impelled to join this movement; their thoughts turned toward the rivers where little factory towns were opening up to make cloth, guns, clocks, buttons, or wire. New machines were making a new life possible to American youth. For the boy who was willing to start at the bottom there was glorious opportunity. The South was being rejuvenated by cotton, a process rendered possible by Whitney's gin and Slater's spinning mills, the North was meeting a demand for mechanized industry, and the West was absorbing as much of its population as the farms and the factories could spare.

Albany stood somewhat apart from these trends because of the manorial tradition introduced by the Dutch, and it was partly because of this backwater condition that Joseph Henry escaped the general influences of his time and was permitted to develop along lines very different from those followed by most American boys of his own circumstances. His failure to display an interest in the work of Doty inclines one to believe that he possessed no aptitude for practical mechanics, and in spite of his association with the soil while at Galway, his interests never wandered toward agriculture.

As a boy he seems to have been as aloof from the general interests of the people as he was to be later as a man. His dalliance with the stage was merely the expression of an ardent nature that had not found its moorings. An abrupt transfiguration of mind was to prove this. That he was frequenting the theatre, making friends there, and was the accepted leader of "The Rostrum" indicated that he was at least a companionable young man.

Lucky circumstances plotted to make Henry a pathfinder. The rabbit which declined to remain tranquilly within its pen led to the chance blossoming of his mind, and a similar accident was to aid in the selection of the channel in which that mind should operate.

A minor ailment kept him indoors. He had to rely for diversion upon the slender household resources, but it happened that one of Mrs. Henry's boarders was a fellow-Scot named Robert Boyle who was able to supplement the few books which the family supplied. Among the books which Boyle placed at Joseph's disposal was Gregory's *Lectures on Experimental Philosophy, Astronomy, and Chemistry*, published in London in 1808.

The author of the book, an English clergyman, caught the attention of the artless reader by a few introductory questions to arouse curiosity. "You throw a stone or shoot an arrow into the air; why does it not go forward in a line with the direction you give it? . . . On the contrary, why does flame or smoke always mount upward, although no force is used to send them in that direction? . . . Again you look into a clear well of water and see your face and figure, as if painted there. Why is this?"

The conversational style of the writer and his simple treatment of scientific subjects captivated Joseph Henry. He had never encountered any book quite like this. It was as though a new casement had opened upon the world.

For the first time the boy was given to understand that there is a meaning behind many familiar phenomena, and that by study and observation it was possible to discover the hidden meanings of things. Here was a field in which imagination, harnassed to the logical faculty he had doubtlessly inherited from his Scottish forebears, could operate in a satisfactory manner. The combination of mental qualities working on science could furnish more permanent results than any amount of play-acting. No longer a languid reader, he began to sample the joys of scientific reading.[2]

He talked eagerly about the book with Boyle, who was so impressed with the boy's enthusiasm that he made him a present of the work. The volume remained a treasured possession long after its contents ceased to have any value to Henry. More than a quarter of a century later, when Professor Joseph Henry had made an international scientific reputation, he came across the old volume in his library. As he turned its pages a flood of memories was stirred. He recalled the impression it had made on his mind at the first reading, and he inscribed upon its flyleaf this grateful tribute:

> This book, by no means a profound work, has under Providence exerted a remarkable influence upon my mind. It accidentally fell into my hands when I was about sixteen years old, and was the first book I ever read with attention. It opened to me a new world of thought and enjoyment; fixed my attention upon the study of nature, and caused me to resolve at the time of reading it that I would immediately devote myself to the acquisition of knowledge.
>
> J. H.

Turning from the futilities of the rhetoricians to follow a livelier oracle, Henry, with a last dramatic flourish, made formal renunciation of his former life in a valedictory address to the members of "The Rostrum."

Once it had been revealed to him that the secrets of nature could become his through the pursuit of knowledge, Henry must have been appalled by his lack of equipment for the quest. He who had so little education must yet have known that a youth was a dunce who was not ready for college at sixteen. However, he had more in him than many a boy of his age who had taken Greek and Latin, and who had had a supply of "head knowledge" flogged into him. Above all, he had the determination to succeed; Scotch blood flowed in his veins, and self-improvement is almost a life principle with the Scot.

---

[2] Faraday was introduced to science in a similar way, by the chance reading of Mrs. Marcet's *Conversations on Chemistry*.

EARLY DAYS

His situation was not unique. Many of the great scientists had been self-taught so far as their science was concerned. The only schools where an adequate training might be obtained were the dissenting academies in England, where men like Dalton taught, and the Artillery Schools in France, which produced Napoleon.

The town of Albany had a good school, the Albany Academy, which Dr. Nott, president of Union College, described as a "college in disguise." Henry resumed his education by attending night classes here in geometry and mechanics. He soon supplemented these studies with a course of English grammar under an itinerant instructor. With the aid of this teacher he acquired a fluency in etymology, syntax, and prosody. He so quickly mastered this subject that before long he judged himself qualified to teach it. In order to earn enough to continue his studies at the Academy, he began to visit outlying districts offering his services as a teacher of grammar.

Acquiring knowledge without a formal curriculum is often a perplexing task. An impending grasp upon one subject almost invariably proves the need for undertaking the study of another. But while his knowledge betrayed the need for extending the base of his studies, the crushing burden of poverty threatened to frustrate all his efforts. The inescapable fact confronted him that he must find an occupation that would provide him with the means for continuing his studies. He applied for a vacant position as teacher in one of the district schools. Although he was judged somewhat young for the post, he was appointed at a monthly salary of eight dollars. He must have done his work well, for at the end of the first month he was given an increase in salary to fifteen dollars.

This position as schoolmaster, which he had obtained by his own efforts, is remarkable for being the only post or appointment for which he ever applied. Thereafter he displayed such ability that others came to him unsought.

During his student days, Joseph Henry was tall and slender, but it was fortunate that the lathy young man had a robust constitution, for between his work and his studies he was compelled to spend about sixteen hours a day in the classroom, either as teacher or as student. He claimed that this was accomplished without sense of fatigue.

At the end of seven months of this program he passed the examination of the Academy with honors. At the time when he took the examinations, and when he was about to begin the study of the calculus, General Stephen van Rensselaer, the local patroon, was in need of a private tutor for his children. As the General was head of the Academy's Board of Trustees, it was natural that he should apply to its principal, Dr. Romeyn Beck, to recommend someone for the post. Beck had been impressed by Henry's earnestness and determination, and gave him such a high recommendation that he was appointed to the vacancy.

Henry was tutor to the van Rensselaer children for two years, guiding them through the Grandisonian curriculum favored by the gentlefolk of the time. The duties were far from arduous, being discharged in about three hours each day. This left him with abundant leisure to follow his own devices, which meant that he was able to do more private study.

A clearer instance of the selection that marks the man of original mind than Henry's insistence upon providing himself with the fundamental knowledge for scientific investigation can scarcely be found. There was nothing in his environment to help. Most of the comforts of life were denied him.

Henry's late intellectual development was probably no disadvantage; Nature's strongest plants are not produced by hot-house forcing. Henry's tenacity and the strength of his desire for an education would be rare in any generation. He learned by his own efforts, and was not merely taught. Meanwhile, he was contemplating how he might apply the knowledge he was set upon acquiring.

Science was not yet a profession. The scientist was still so much of an amateur that a Cambridge professor declared it to be "a proper occupation for the leisure of an English gentleman." Dr. Beck could have enlightened Henry upon the barriers facing the man who wished to adopt a scientific career. Some of these difficulties were enumerated by John Torrey, when he was following the same course as Henry. "Inaccurate and undependable works by previous authors, a lack of patronage in science, jealousy among the members of the science, stinginess of materials, lack of ample library facilities, these and many other hardships rose like phantoms to discourage more than one young scientist of less strong stuff." [3]

Except teaching in a few colleges like Harvard, Pennsylvania, the College of New Jersey, and, more recently, Yale, the only profession open to an aspirant in science was that of medicine. Having no alter-

[3] Rodgers. *John Torrey*, p. 28.

native, Henry began to apply himself seriously to the study of "physik" by studying anatomy under a Dr. Tully and physiology under Dr. March. In October 1825 he also began to assist Dr. Beck in preparing his chemical lessons and demonstrations. The association with Romeyn Beck was both pleasant and profitable. Amos Eaton, who knew the three Beck brothers, described Romeyn in a letter to Torrey as "though a pretty good fellow, is given to too much puffing." [4] But to Romeyn Beck's credit stands the fact that he recognized Joseph Henry's outstanding abilities from an early stage. He never ceased to offer his student encouragement and opportunity. A younger brother, Lewis Beck, the local botanist and a member of the Academy faculty, was also a close friend of Henry, although it is doubtful whether he exercised the same influence as Romeyn.

It is likely that the egotistical and sometimes erratic Amos Eaton was introduced to Henry by the Becks. Eaton had been sentenced to a term of imprisonment on an ambiguous charge of embezzlement, but had been unconditionally pardoned. He possessed the confidence of influential men like Governors Clinton and Marcy, and General Stephen van Rensselaer. Besides being a geologist and botanist, Eaton was a capable chemist and was to introduce the laboratory method of teaching chemistry into America one year before Liebig opened his laboratory at Giessen, which attracted so many young American students.

Eaton had a host of friends, many of whom must have visited him in the "old Dutch gambril-roof brick house behind an elm tree" in Albany where he settled in 1818 after leading a Tartar's life as a peripatetic teacher of geology and chemistry. Although the ostensible purpose of his stay was to see a book through the press, Eaton was too energetic to remain without further occupation. He decided to give a series of lectures on botany which Henry, with a medical career in view, may have attended, for the doctors in those days had to possess a knowledge of plants and herbs in order to make their own medicines.

Friendship with Eaton proved very beneficial to Henry. Apart from his knowledge of chemistry, geology, and botany, he had many other interests, and he had a rare faculty of being able to communicate his enthusiasms to other men. It seems certain that Eaton was drawn to the younger man by Henry's attainments; in 1826, shortly

[4] McAlister. *Amos Eaton*, p. 248.

after the Rensselaer School opened in Troy, he recommended that Henry be appointed one of the school's examiners in mathematics. Beyond Eaton and the Becks there were few friends who could give young Henry practical aid in his scientific studies, but to these he was indebted at this early stage.

At this time, attendance at philosophical and scientific lectures had become a fashionable pastime, and Romeyn Beck's chemical lectures drew large audiences to the Albany Academy. This might have enabled his assistant, Joseph Henry, to extend his social acquaintances, for he had to arrange Beck's apparatus. He had already impressed his master as an unusual experimenter on his own account. No sooner was an interesting experiment reported by some other investigator, than Henry set up the apparatus to repeat it. Having thus, to borrow Goldsmith's expression, "travelled over other men's minds," he would proceed to test the principle involved and, if possible, to extend its range and operation. His knowledge of chemistry was not sufficient to lead him to any noteworthy result, although he proved adept at improving the demonstrations and experiments.

The city had more than its usual quota of educated people. As the capital of the State it naturally attracted a large professional population, but those who engaged in legislative and legal affairs did not account wholly for the intellectual activity of the city. Like many of the other towns and cities of America, its people were making tentative efforts to acquire a knowledge of the broader world in which they lived. The intellectual ferment found expression in the meetings of the numerous clubs and societies that were formed to discuss literary and scientific trends. Albany was not behind in this social movement. It soon had two scientific societies.

Such a community was not by any means a poor place for a budding scientist to make his debut. If the audience is sufficiently intelligent to be appreciative and encouraging, on a smaller stage merit has a better chance of recognition. In thinking of Henry's association with Albany Academy it is well to recall that Sir Humphrey Davy had a similar scientific apprenticeship in the Medical Pneumatic School at Clifton before he went to the Royal Institution. Dr. Olando Meads gives an interesting account of Henry in these days in an address read before the Albany Institute:

"When a boy in the Albany Academy in 1823-4, it was my pleasure and privilege, when released from recitations, to resort to the chemical laboratory and lecture room. There might be found from day to day through the winter, earnestly engaged in experiments upon steam and upon a small steam engine, and in chemical and other scientific investigations, two young men, both active members of the "Lyceum," then very different in their external circumstances and prospects in life; the one was Richard Varick DeWitt [5] the other was Joseph Henry, as yet unknown to fame, but already giving promise of those rare qualities of mind and character which have since raised him to the very first rank among experimental philosophers of his time.

"Chemistry was at that time exciting very great interest, and Dr. Beck's courses of chemical lectures, conducted every winter in the lecture room of the Academy, were attended not only by the students, but by all that was most intelligent and fashionable in the city. Henry, who had been formerly a pupil in the Academy, was then Dr. Beck's chemical assistant, and already an admirable experimentalist, and he availed himself to the utmost of the advantages thus offered, of prosecuting his investigations in chemistry, electricity, and galvanism." [6]

In spite of public curiosity about nature and interest in promoting the useful arts, Albany was not populous enough to support the two scientific societies which its citizens had started. Nor were there sufficient experienced men to serve as leaders for two groups. An amalgamation was in order. This was brought about in 1824, when the Albany Institute was formed by the junction of the Lyceum and the Society for the Promotion of the Useful Arts, with about two hundred and fifty active members under the presidency of General Stephen van Rensselaer.

Joseph Henry was appointed librarian of the Institute, which soon had a collection of three hundred volumes, a small library as modern standards go, but a very respectable collection in those days. Henry's gifts to the library and its collections of specimens form a strange miscellany. In 1824 he gave a collection of lithographic printings on satin. In the following year he presented the society with two copies of Genet's *Memorial on the Upward Forces of the Fluids,* and to the specimen cabinet he and Dr. Elial T. Foote donated "several speci-

---

[5] Son of Simeon DeWitt, Chancellor of Board of Regents of the University of the State of New York.
[6] *Transactions of the Albany Institute,* 1872. Vol. VII, p. 20.

mens of native carburetted hydrogen gas from Fredonia and Portland, Chautauqua County. . . . On passing it through the apparatus used in burning artificial gas, it was found to yield a beautiful clear light, corresponding in color with the purest forms of manufactured carburetted hydrogen." [7] This gas was probably obtained on one of the surveying excursions he had made with Amos Eaton. In 1827 he gave a specimen of *Strombus pugilis,* a species of the winged shell of the *Strombidae* family.

Far too modest about his own attainments to have volunteered for the task, Henry was persuaded to address the members of the Institute on the experiments he had performed with the model steam engine. His first communication was read on October 30, 1824, when he was twenty-six years old. His subject was entitled "On the Chemical and Mechanical Effects of Steam; with experiments designed to illustrate the great reduction of temperature in steam of high elasticity when suddenly expanded." [8]

The paper with the formidable title dealt with the elementary principles of what is now known as "adiabatic expansion." Neither Henry nor any of his listeners could have had the slightest conception of the uses to which the expansion of steam would be applied, yet he contrived to give his audience an impressive demonstration. By holding a thermometer in jets of steam allowed to escape from a boiler under varying internal temperatures, he demonstrated how the temperature of a jet of steam diminished as the temperature and pressure within the boiler increased. Then, he showed that a jet of escaping steam would not scald his hand when exposed to it, if the original temperature of the steam was sufficiently high.

Perhaps if any significance is to be found in this paper it resides in Henry's interest in the growth of industrial techniques. Steam was slowly making its way into industry, and there was already talk of the "steam wagons" which were to supersede all other forms of land transportation. There was much talk about steam, although little was known about it, so that Henry's audience must have felt that they were being admitted to its inner mysteries.

The first paper having proved a success, he was urged to read another. This was delivered on March 2, 1825, and his subject was "The

[7] Albany Inst. *Trans.* Vol. I, p. 53.
[8] The paper is not printed in the *Transactions* of the Institute, and I have had to depend upon the account of its contents given by W. B. Taylor in "The Scientific Work of Joseph Henry." *Memorial,* p. 209.

Production of Cold by the Rarefaction of Air." As in all his lectures he devoted especial attention to the demonstrations, one of which on this occasion was noteworthy.

One of these illustrations most strikingly illustrated the great reduction of temperature which takes place on the sudden rarefaction of condensed air. Half a pint of water was poured into a strong copper vessel of a globular form, and having a capacity of five gallons; a tube of one fourth of an inch caliber with a number of holes near the lower end, and a stopper attached to the other extremity, was firmly screwed into the neck of the vessel; the lower end of the tube was dipped into the water, but a number of the holes were above the surface of the liquid, so that a jet of air mingled with water might be drawn from the fountain. The apparatus was then charged with condensed air, by means of a powerful condensing pump, until the pressure was estimated at nine atmospheres. During the condensation the vessel became sensibly warm. After suffering the apparatus to cool down to the temperature of the room, the stopcock was opened; the air rushed out with great violence, carrying with it a quantity of water, which was instantly converted into snow. After a few seconds the tube became filled with ice, which almost entirely stopped the current of air. The neck of the vessel was then partially unscrewed, so as to allow the condensed air to rush out around the sides of the screw; in this state the temperature of the whole interior atmosphere was so much reduced as to freeze the remaining water in the vessel.[9]

No new theory was involved; indeed, it was only a more spectacular demonstration of the same law he had employed in his previous paper, but the younger members of the audience lived long enough to see the principle applied in commercial ice-making. Perhaps the most interesting feature about this demonstration was the unfavorable conditions under which it was performed. The night was unseasonably warm, the room was heated by a large stove situated near the lecturer, and a large audience filled the hall. This combination of circumstances drove the room temperature to nearly 80 degrees, a condition which caused Henry no small amount of perturbation over the success of the demonstration, but which must have greatly im-

[9] Albany Inst. *Trans.* Vol. I, pt. 2, p. 36.

pressed those members of the audience who realized his predicament.

Meanwhile, he was not neglecting the study of mathematics. He had now progressed to Lagrange's *Mécanique Analytique,* that great treatise in which the Frenchman so richly illuminated the study of mechanics, in which his differential equations gave to the subject a new generality and completeness. But Henry never became a mathematician.

Somehow, while engaged with his studies and while he assisted Dr. Beck, he found time to act as private tutor to Henry James, the lad who was to become the father of two famous sons, William James, the psychologist, and Henry James, the novelist. Joseph Henry was supposed to have been supervising a group of boys during the recreation period. As was a common practice, the boys were engaged in a game in which a ball soaked in turpentine figured. This ball was then set on fire. Through some mischance, the flaming ball set a hayloft on fire. Young James bravely beat out the flames with his feet before much damage was done, but in the action he suffered such burns that he had to have a leg amputated. He was unable to follow his studies at the Academy as a result of this injury, and was prepared for college by Joseph Henry. With some interruptions, the association of tutor and pupil was maintained after James had gone to Princeton.[10]

But Nature, no patron of genius, was exacting her toll for the strain to which Henry was subjecting himself in his studies. The ceaseless grind kept him too much indoors, which proved detrimental to his health. When it became apparent that he must relax his efforts, he was unexpectedly offered, through the influential Judge Conkling, a position as surveyor upon a projected new road between West Point and Lake Erie. The proposal of self-sustaining labor in a healthy outdoor occupation was exactly what Henry needed.

The route he had to follow extended from Kingston, near West Point, to Portland Harbor on Lake Erie. He threw himself wholeheartedly into the task, nor was it any easy task when winter came and the line had to be followed unwaveringly in deep snow over wild and rugged country, when the wind was sharp and cold. Nevertheless, he enjoyed the work and completed it with credit. Indeed, he had done so well that his friends sought to secure for him a place in the U.S. Corps of Civil Engineers, but it had been determined that all future vacancies in this Corps should be filled by West Point

[10] Rukeyser, M. *Willard Gibbs,* p. 109.

graduates. Mr. John D. Hammond, who wrote in recommendation of Henry's application, said he did not know any person "more free from faults or more amiable in his disposition."

The work in the vigorous air had done more than bring Henry a reputation. Rediscovering solitude and examining the distance had relieved his mind from tight pressure and had permitted it to expand. His health, which had suffered from sedentary occupations, was completely restored. Thereafter he was to be distinguished for a ruggedness of constitution which endured to the last day of his life. It was notorious that he could work for long periods of acute concentration which would have left most men exhausted, but which left him apparently unaffected. He never showed signs of fatigue. He did not share in any degree his father's delicate health, but seems to have skipped a generation back to draw a haleness from a hardier type of Scot.

The failure to secure the Federal post as a civil engineer was balanced by other offers of employment to supervise the construction of a canal in Ohio and to manage a mine in Mexico. These were alluring offers and he was on the verge of accepting the former in the spring of 1826, when Dr. Beck informed him that the chair of Mathematics and Natural Philosophy at the Albany Academy would soon be vacant and that Henry would be appointed to fill the post.

For the first time Henry, who had known only struggle and stress, was torn between two pleasant prospects. To arrive at a position on the staff of the Academy must have seemed to be the summit of his ambition, yet he was under a strong inclination to accept the post in Ohio. Once a man has tasted the joys of working in the open air it is hard to relinquish the sense of freedom it affords, but the earnest solicitation of his friend and mentor won him over to acceptance of the position at the Academy, although it meant taking the lower remuneration.

Henry was never greatly influenced by money, as we shall see later. This brief spell of occupation as a surveyor was the only work he ever had that was profitable enough to enable him to save. He received from the State of New York the sum of $2,083 for his survey work, and the sum he was able to save from this amount was placed in the hands of a friend for investment. The interest was rarely drawn but was added to the capital; this small sum, together with an insurance policy for $4,000, represented his life's savings.

WORK AT ALBANY

In 1826 the nation was suffering from growing pains. It was the height of the turnpike era, when those countless roads over which we travel so briskly were coming into existence. The State of New York had built fifteen hundred miles of stone and gravel roads. The Federal government had authorized construction of the famous Cumberland Road which served as a backbone to the exodus of civilization over the mountains. But already the transition was setting in which changed the pattern of transportation.

The Erie Canal and other water arteries were being cut as quickly as funds and qualified men could be found. Fulton had succeeded where Fitch and others had failed to convince the public that the steam engine was designed to solve the problem of travel by water. By the year 1830 the Federal government had withdrawn support for road-making and was financing river, harbor, and lake improvements to promote the flow of traffic and passengers. The Stevens family, Peter Cooper, and Oliver Evans were talking of putting the steam engine on wheels and running it on rails.

The churning of ideas over this expanding transportation inflamed hopes of increased prosperity, and there were opportunities innumerable for men with the abilities which Joseph Henry had recently displayed. The future seemed to lie out-of-doors, in the expansion of industry and commerce suddenly rendered possible by the introduction of the steam engine. Few young men with ambition could withstand the attraction. It must have tugged strongly at Henry's inclination.

Why should Henry have cast aside these opportunities in order to become a schoolteacher with such narrow prospects? At this time his two most valued friends, upon whose counsel and riper experience he most relied, were Eaton and Beck, both educators. They could point out to him that he had already proved himself a good instructor and that the nation had need of teachers as it never had before. Ambition might pull in one direction but duty pulled with greater force in the opposite direction. The choice cannot have been easy. At other times of his life, as we shall see, when Henry had a similar decision to make he took the same course and followed the path of what he conceived to be his duty. He did not act blindly. He was conscious of the sacrifice his choice imposed, but apparently he achieved greater satisfaction from following the path of duty than from worldly advancement.

The decision to become a professor meant that he had to renounce any intention to become a medical doctor. It is impossible to say whether Henry would have become a great or even a good doctor, or would have found happiness in that profession. Probably not. He was not lacking in human sympathies, but the living organism did not interest him to the same extent as did inanimate objects. He wandered far in the various sciences, into meteorology, mineralogy, geology, but there is no evidence that he was ever afterward to study botany or zoology or living man.

Moreover, it is not improbable that with his expanded knowledge of natural science, Henry may have felt a slight resentment against the medical doctors for their attitude to the subjects which held a fascination for him. The doctors were regarded as the representatives of science and, while medicine enjoyed an immense prestige, its practitioners considered their other studies, such as chemistry and physics, socially inferior. Henry did not share this view. These sciences had their own dignity, and at the beginning of the nineteenth century were acquiring a great influence, especially chemistry, which had made rapid advances.

Henry received his appointment as Professor of Mathematics and Natural Philosophy in April 1826, but since his duties did not begin until September he had nearly five months in which to prepare himself for the work. The time was not spent in idleness.

A portion of the period was spent in making a geological and geographical survey of the surrounding counties under the supervision of his friend Eaton. No direct account of this expedition is preserved in Henry's scientific writings, but on October 28, 1829, he read a paper before the Albany Institute entitled "A Topographical Sketch of the State of New York"[1] which had been written as an introduction to the atlas of the State published by David H. Burr.

The article is largely concerned with the surveys Henry had made, together with those made by others, and is accompanied by numerous tables giving the mileages of various routes followed, and elevations. Here he gives us a glimpse of the pride he took in attaining accuracy in this work, for he wrote: "Table No. 2 also intersects with Table No. 4 at Binghampton, and with No. 7 at Newtown. At the former place the difference was only the fraction of a foot, and at the latter place less than two feet. These facts show with what precision measurements of this kind can be made, and what reliance can be placed on the correctness of the elevations of the several points given in these tables."

It is conceivable that during this period of intimacy with Eaton, Henry sought to improve his ideas on education. Eaton had been appointed Senior Professor at General van Rensselaer's School of Science[2] in 1824, and was now a figure of considerable importance in the educational world. He held enlightened views on the teaching of science and passed as an advanced liberal in that respect. Many of Eaton's innovations have since been widely accepted.

During the summer Henry managed to organize and plan his courses for the coming school sessions. The course he devised was to comprise the very latest principles in natural philosophy, and the principles were to be demonstrated liberally by illustrative experiments. This was an innovation. Most teachers were satisfied to work with a text-book and a blackboard. That Henry's methods found favor with his students is attested by the ease with which the instructor was able to find witnesses who remembered these demonstrations, and who sprang eagerly to Henry's defense when he had to safeguard his integrity in the famous telegraph controversy.

He was now at the threshold of the scientific career which was to bring him distinction. He had been working toward medicine and had been diverted into civil engineering and to geology. The diversity would suggest that he had been drifting with the current, but we must

---

[1] Albany Inst. *Trans.* Vol. I, p. 87; *Scientific Writings.* Vol. I, p. 8. In references and footnotes of this book *The Scientific Writings of Joseph Henry* are abbreviated *S.W.*
[2] Later, the Rensselaer Polytechnic Institute.

remember that he had a purpose. He was attempting to expand his knowledge spherically, not all in one direction. A hard student, capable of great powers of application, he was accumulating a store of knowledge upon which he could draw as occasion demanded. Throughout his life he never ceased to extend his knowledge as new subjects were presented.

He entered upon his academic duties in September 1826. His appearance as a member of the staff of such a highly reputable institution evidently aroused a feling of family pride among his relatives for it prompted one to write him this congratulatory letter:

Rochester
February 11, 1827

Cousin Henry:

Surprise is one source of pleasure, an effect this letter may possibly produce since the long silence between you and me has never been interrupted by any paper communication. All the merit these lines can claim will be perhaps to produce an echo.

I write to tell you I have seen an extract from your inaugural address and the only reason I was not more delighted was that I could not see the whole of the production which is worthy of the Professor, a feast to the men of intellect as well as to the man of taste. From what I could gather from the *Argus* I must congratulate you on its success at the time of its delivery before a learned and polite auditory. May your professorship be as splendid as its commencement was brilliant. . . . Let me exhort you not to give yourself exclusively to the dry bones of diagrams but consider that you were partly made for your friends and social intercourse.

Yours most affectionately,
Alexander S. Alexander

The Albany Academy is situated upon a public square near the head of State Street, close to the Capitol building. The building is maintained very nearly in its original condition by the State of New York, and is called the Joseph Henry Memorial. The educational activities center in a new building.

In Henry's day it was not a large school, with its four professors and one hundred and fifty students. Still, there were few colleges in the country offering a wider course of study in mathematics and

WORK AT ALBANY

natural philosophy. The professor who covered these subjects had to be a man of parts. Henry not only fulfilled the requirements of the post in respect to breadth and variety, but he displayed an uncommon degree of modernity in the introduction of experiments to illustrate the lessons. His theatrical experience undoubtedly helped him in his classroom work. He possessed an easy and impressive manner of lecturing and a simple but forceful method of expression. He kept a copy of Johnson's dictionary on his desk, and any word that could not be found in it was banished from his vocabulary.

To a man who had just completed twelve months of outdoor occupation, the vista of an endless procession of days within a classroom could have presented an aspect of unrelieved drudgery, a mere means of eking out existence. For seven hours a day he had to teach mathematics, physics, and a little chemistry, one half of that time being devoted to teaching the smaller boys the rudiments of arithmetic. He uttered no word of complaint against the tedium of the task, but we do no violence to his gentle nature if we suppose he welcomed the hours given to more advanced pupils who could share his enthusiasms.

The routine of schoolwork did not absorb all his energies. Although he was still acting as private tutor to Henry James, there was a margin of time that could be devoted to the investigation of new scientific problems.

The weather is always an acute problem to an agricultural and seafaring community. In the third decade of the nineteenth century, some of the states were making tentative efforts to benefit from the organization of societies by encouraging the members to collect reliable information about the weather, which would replace dependence upon old saws and superstitions relating to the behavior of birds and domestic animals as forecasting material.

The Regents of the University of the State of New York, who supervised the educational establishments, promoted a system of meteorological observations by distributing a number of thermometers and wind gauges. The observers were requested to devote special attention to the study of weather conditions on those days when the behavior of domestic animals and wild birds showed a departure from the normal. Having enlisted a corps of observers, the Regents soon found themselves burdened with a mass of unrelated data.

In 1827 Simeon de Witt, chancellor of the Board of Regents, concluded that this material required tabulation and correlation. He engaged Dr. Beck to undertake the work, and Beck recruited the aid of Joseph Henry. Henry plunged into the work with zest. He thus acquired an interest in meteorology which never left him.

The first abstract of these collections, for the year 1828,[3] comprised a tabulation of the readings and observations of the monthly and annual means of temperature, wind, rainfall, etc.; a general account of meteorological incidents, and a table of "Miscellaneous Observations" on the data of notable phases of animal and plant phenomena associated with climatic conditions.

The compilation of these annual reports of the Regents must have consumed no inconsiderable amount of Henry's leisure time. He embellished the third (for 1830) with a careful calculation of the latitude, longitude, and altitude of each of the State observation stations. This was the gratuitous contribution of a student with the ability to perceive that a mere factual table lacked the essential factors for interpretation.

Meanwhile, since his student days, Henry had been interested in electricity and magnetism. During the same period that he was working on the meteorological data, he was conducting investigations in electromagnetism that were to give him international fame.

In order to understand and appreciate the contributions of Joseph Henry to the science of electricity and magnetism, it is necessary to examine the state of knowledge at the time he did his work. When Henry began his study of electricity, the science was enjoying one of its periods of popularity. In its early stages electricity had the quality of advancing under the impulse of some bold philosopher's discovery and then subsiding abruptly when a grand idea was bequeathed to inferior minds. Progress was checked until, by impulse or by accident, an unusual experimenter was enabled to revitalize interest by another advance.

Henry may be regarded as fortunate that he should have entered upon his studies at a time when knowledge of electricity was superficial. At the opening of the nineteenth century very little was known about electricity or magnetism and, while a correlation was occasionally suspected by some acute thinker, it was not ever definitely known that electricity and magnetism were related phenomena. Both mag-

[3] *Annual Report of the Regents of the University to the Legislature of New York.* Albany. Vol. I, 1829–1835.

netic and electric phenomena had been known from antiquity. For approximately two thousand years no progress had been made in the understanding of these phenomena. In the year 1600 William Gilbert published the first scientific treatise on the subject.[4] In this great work, old wives' tales were cast aside and magnetism was examined from a scientific viewpoint. Gilbert sought to explore the laws governing the lodestone. He concluded that the earth was a magnetized globe. In extending his investigations, Gilbert found that many substances other than jet and amber possessed the property of attraction through friction. He described this property as electric.

Seventy years were to pass before Otto von Guericke invented the first electrostatic machine, which enabled him to extend the study of electricity. He observed that electricity not only behaved like magnetism in attracting light bodies, but that it also repulsed certain charged bodies. Repulsion was a property which Gilbert had denied to electricity. Guericke also discovered electrical conduction. Unfortunately his work [5] was not widely read, and most of what he had uncovered had to be rediscovered.

Progress quickened in the next century. Stephen Gray pointed out the difference between conductors and nonconductors, and transmitted current through several hundred feet of hempen thread. Meanwhile, C. F. Dufay in Paris was independently engaged on experiments which led him to affirm that there were two kinds of electricity which behaved in opposite ways.

As the friction electric machine was improved and became more widely known, electrical experiments began to attract a wider interest. Electricity became available in larger quantities, and efforts to store it resulted in the invention of the Leyden jar. By means of this intensifying jar, great interest was aroused in the physical shocks electricity could administer, and in its spark effects.

The Abbé Nollet, also working in France, demonstrated the strength of the charge which could be stored in these jars. Henry was probably a close student of his work for he owned a copy of Nollet's book,[6] which he presented to the library of the Albany Institute.

The next great name in electrical science is that of Benjamin Franklin. It was Franklin's good fortune to reside in the dry atmos-

---

[4] *De Magnete, Magnetisque Corporibus, etc.* London, 1600.
[5] *Experimenta nova ut vocantur, etc.* Amsterdam, 1672.
[6] *Leçons de Physique Experimentale.* Paris, 1759.

phere of Philadelphia, which gave him and his associates exceptional opportunity to exercise their skill with the electrostatic machine. As a result, many of their experiments were of an original character.

The famous kite experiment enabled the Philadelphia group to establish what had been surmised by others, that lightning was identical with the mild charges of electricity produced by the friction of glass and silk. Franklin invented the lightning rod, which goes down in history as the first practical electrical invention. The lightning rod removed lightning from the realm of the supernatural and brought it within human understanding. Franklin's greatest single contribution to electrical theory was his "single fluid theory," in contrast to Dufay's "two fluid theory." Franklin explained "positive" and "negative" charges by the presence or absence of a single electrical fluid, much as we do today.[7] This facilitated understanding, since the single fluid explanations were much simpler that those based on two fluids.

Henry was probably well acquainted with these experiments and theories, for it is almost certain he would read Joseph Priestley's book, *History and Present State of Electricity,* which had been published in 1767. This remarkable book was available in Albany, in the State Library, after 1824, and it would be unlikely he would overlook such a prize. The ideas presented in Priestley's work were the best current beliefs at the time of publication.

The contents of this book almost marked the limits to which electrical investigation could be carried without benefit of additional information which could be derived only from a knowledge of electrical circuits. Only one important advance in static electricity was made between the publication of Priestley's book and the opening of Henry's investigations. This was Charles Coulomb's experimental proof of Priestley's deduction that the force between charged bodies was inversely proportional to the square of the distance between them. Coulomb's experiments are of interest because they mark the pioneer attempt at the introduction of measurement, of quantitative observation, in the study of electrical phenomena.

No sooner had Coulomb rendered the phenomena of static electricity comprehensible in a manner satisfactory to the time, than the

---

[7] Note that we now know that positive and negative charges are caused by the flow of negatively-charged electrons, whereas Franklin considered the flow of a positive "fluid." This is the origin of the plus-to-minus flow convention still used in most electrical engineering, though the convention is contrary to modern knowledge.

discoveries of Galvani and Volta opened a new era in investigation.

The contraction of living muscle under the impulse of electric shocks had been known from the time of Guericke. In 1752 Swammerdam, the Dutch naturalist, had observed that when the muscle of a frog is laid bare, a tendon grasped in one hand, and a nerve touched with a scalpel held in the other hand, a muscular twitching would result.[8] He did not follow up this observation.

In 1786 the Italian anatomist Luigi Galvani was studying the effects of electricity on animal organisms when he chanced to make an observation not unlike that of Swammerdam but he made it the subject of a painstaking investigation. Galvani describes the discovery he made in a few simple words:

> I had dissected and prepared a frog . . . and while I was attending to something else, I laid it on a table on which stood an electrical machine at some distance. . . . When one of the persons who were present touched accidentally and lightly the inner crural nerves with the point of a scalpel all the muscles of the legs seemed to contract again and again. . . . Another one who was there, helping us in the electrical researches, thought he had noticed that the action was excited when a spark was discharged from the conductor of the machine.[9]

The essential feature of Galvani's subsequent studies was the discovery that the electrostatic machine could be dispensed with if the leg and the nerve were connected by two *different* metals. When he found that outside electrical disturbance was unnecessary he concluded that electricity must be present in the animal itself. He imagined he had discovered the vital fluid.

His observations produced an excitement throughout the scientific world fully equal to that produced by the invention of the Leyden jar. Other workers hastened to repeat his experiments. Among those who investigated Galvani's work was Alessandro Volta, an Italian physicist. He was the first to perceive the real nature of his compatriot's discovery. In 1796 Volta eliminated the organisms and succeeded in making electrification possible by bringing two metals into contact.

This was a fundamental discovery of prime importance. From it

---

[8] Swammerdam, J. *Biblia Naturae.* Leipzig, 1752. Vol. II. Quoted in Mottelay.
[9] Galvani, L. *Opere: edite ed inedite.* Bologna, 1841, p. 63.

sprang the Voltaic pile, which was disclosed in a letter to the president of the Royal Society in London.[10]

Volta announced that he could produce continuous electricity by erecting a pile of zinc and copper discs placed alternately one on top of the other, with an intervening disc of moistened paper. The combination of the two metals was the formation of the first primary battery, each pair of discs separated by damp paper formed a cell, and gave a difference of electrical potential. The differences of the individual cells when added together produced a marked difference between the potentials of the zinc and copper discs at the respective terminals of the pile.

Volta introduced a variant to the pile in a "crown of cups," which comprised a series of cups filled with brine or dilute acid, each containing a pair of plates, one zinc and the other copper. The zinc of one cup was connected to the copper of the next, until isolated copper and zinc plates were left in the first and last cups. Current was drawn from these terminal plates.

Although this was a clumsy arrangement, and the intensity remained weak unless a large number of cups was brought into use, Volta's discovery of the electric battery provided a source of continuous current, in contrast with the momentary spark discharges from the Leyden jar. This distinct advantage ensured for the battery an attention that was bound to result in improvements. The first of these was made in England when William Cruickshanks constructed a trough with grooved sides and bottom. Into the grooves he inserted plates of copper and zinc soldered together so that all zinc surfaces faced one way and all copper surfaces faced in the opposite direction. Cruickshanks' trough was converted into a primary battery when the cells formed by these plates were filled with dilute acid. This trough did not eliminate one defect of Volta's pile, the electrolytic polarization created when hydrogen bubbles formed on the copper plates. Nevertheless, the Cruickshanks trough was to figure in some famous experiments.

The scientific journals were filled with accounts of the marvels performed with the aid of Volta's discovery. The first fruits were harvested by the chemists. Progress outside chemistry was not reported until 1802, when an Italian jurist named Gian Dominico Romagnosi announced another remarkable advance, which passed almost un-

[10] Royal Society. *Transactions,* 1800, p. 403.

noticed because he chose to announce it through the medium of a little known newspaper.[11] The substance of Romagnosi's discovery was that a galvanic current would deflect a magnetized needle.

The scientific mind was not prepared for this evidence of electromagnetism. It was felt that the limit of apprehension of electricity had been reached in the discoveries of Galvani and Volta. This complacent attitude, combined with the obscurity of Romagnosi's announcement permitted a discovery of first magnitude to be ignored until the Danish scientist Hans Christian Oersted rediscovered it twenty years later. It has been hinted that Oersted had knowledge of Romagnosi's prior discovery, for it had not passed entirely unnoticed,[12] but probably Oersted's discovery was entirely accidental. No one appears to know what experiment Oersted was performing at the time, but the discoverer's credit is not diminished by his not knowing beforehand what he will discover. It was no chance that placed a Voltaic pile and a magnetized needle on Oersted's table. The same apparatus must have appeared on many laboratory tables before the needle moved under eyes that could understand its significance.

Oersted confirmed Romagnosi's discovery that when a wire carrying an electric current was laid parallel to a magnetized needle, the needle would be deflected. By further experiment he found the relation between the conventional direction of the current, the position of the needle, and the direction of deflection. His announcement of these experiments[13] was the first important step in the science of electromagnetism.

The scientific world might ignore, even if it were aware of, the statement of an unknown Italian lawyer, but when the statement had the support of the professor of physics at the University of Copenhagen, it set the scientists afire with curiosity to know more.

Oersted's discovery was extended swiftly and brilliantly by the French philosopher and mathematician, André Ampère, who conducted a notable series of investigations into the mutual attractions and repulsions existing between straight wires through which elec-

---

[11] *Gazetta di Trentino*, Aug. 3, 1802. A translation will be found in Fahie's *History of the Electric Telegraph*, p. 259.

[12] It was mentioned in Aldini's *Essai theoretique et experimentale sur la Galvanisme*. Paris, 1804, p. 191; and in Izarn's *Manuel du Galvanisme*. Paris, 1805, p. 120.

[13] *Experimenta circa affectum conflictus electrici in acum magneticam*. Copenhagen, 1820.

trical currents were flowing. Ampère also carried out experiments with wires bent into circular form, and with coils composed of many circular windings which he called "solenoids." He compared a magnetized rod to a coil carrying a current. He was able to show that two current-carrying coils act upon one another through their ends in exactly the same way as do magnetized rods through their poles. Going further, he showed that a single circular current, when suspended so as to be free to move, sets itself like a magnetized needle with reference to the earth's magnetism.

According to Ampère, every magnet should be imagined as composed of circular currents similar in direction and set side by side, and could actually be substituted in all effects of its forces by such a circular current. It was only necessary to assume that every iron or steel molecule of the magnet contains in itself a small permanent circular current. Magnetization would then consist of directing all these circular currents of the innumerable molecules in the same sense.

This hypothesis did not meet with a ready acceptance. Although Newton's authority supported the general idea of the atomic constitution of matter, and Dalton had already published his classic views on the subject, few physicists were willing to attribute to such tiny particles the unusual properties Ampère ascribed to them. The objection was raised that the continuance of such currents without a source of power was not possible. This objection was not wholly overcome until a much later date.

Not even Ampère's proofs that the forces due to electric currents, which he had reduced to the law of inverse squares and had thereby brought into line with gravitation, weakened the objections. Henry himself rejected Ampère's brilliant theory upon typical Newtonian grounds. Speaking upon the subject of the "imponderables," he remarked, "The theory of Ampère, though an admirable expression of a generalization of the phenomena of electromagnetism, is wanting in that strict analogy with known mechanical actions which is desirable in a theory intended to explain phenomena of this kind." [14]

The physicists, or natural philosophers, were self-confident. All effects, they thought, could be reduced and explained by Newton's laws of mechanics. The universe was a great machine; smaller entities

[14] S.W. Vol. I, p. 305.

were merely smaller machines, but mechanical, none the less. It explained everything physical.

Dominique François Arago noticed that the connecting wire of Volta's pile would coat itself with iron filings which, although previously entirely nonmagnetic, would afterwards show signs of permanent magnetism. When he brought this to the attention of his friend Ampère, the latter instantly proclaimed that a current-carrying coil would turn steel needles placed in its axis into magnets, having north and south poles arranged in a manner which could be predicted. This opened the way for Sturgeon to make an electromagnet, which gained or lost its power by closing or opening the current-carrying circuit. All he did was to replace steel by soft iron in the core.

Although the first three decades of the nineteenth century were highly fruitful in experiment, only one other discovery of the period calls for mention. Johann Schweigger of Halle had ascertained that the deflection of a magnetic needle produced by the outward current of a primary battery, flowing over the needle, as in Oersted's experiment, had the same effect as the return current flowing under the needle. By leading his wire from a battery above the needle and returning it underneath, he doubled the deflection of the needle. By giving the wire an additional turn the effect was further augmented. A rough measuring instrument was thereby constructed. Henry called it "Schweigger's multiplier." The modern name, galvanometer, is due to Ampère.

Some seventy-five years before Henry was appointed professor at Albany Academy, Benjamin Franklin and a group of friends had performed the series of experiments which marked a new era in the science of electricity. For some inexplicable reason this group did not inspire any band of disciples to pursue their work. Simon Newcomb, in drawing attention to the void left by Franklin, made this interesting comment:

> It is one of the most curious features in the intellectual history of our country, that after producing such a man as Franklin it found no successor to him in the field of science for half a century after his work was done. There had been without doubt plenty of professors of eminent attainments who had amused themselves and instructed their pupils by physical experiments. But in the

department of electricity, that in which Franklin took such a prominent position, it may be doubted whether they enumerated a single generalization which will enter into the history of science.[15]

The problem raised belongs to the sociologist rather than to the scientist. The experiment with liberty, the inter-relationships between men within the nation, pioneering, the slow emancipation from European sources of knowledge, largely account for the vacuum.

One feature which merits more than passing attention was the lack of speedy communication and the absence of scientific publications. Nothing is so difficult for a student to understand as existence in an America where communication was deplorably slow. Our own age is so competently articulate, the transmission of thought is so nearly instantaneous, that an effort is needed to comprehend the slow and silent world in which Henry lived.

Dr. Robert Hare of the University of Pennsylvania, best known for his invention of the oxyhydrogen blowpipe, was the most prominent American in the study of galvanism. Henry was familiar with his work, for it had been described in Silliman's *American Journal of Science,* the first important scientific journal in this country. Hare had invented a "deflagrator," which Henry used at Albany and Princeton. This was a battery so arranged that the plates could be removed easily from the electrolyte to disperse the bubbles of gas which caused polarization.

A few others, like Dana of Dartmouth, Butts of Maryland, and Eaton, now at Troy, were covering the same ground as Hare. But no progress was being made. These men were doing no more than to repeat the experiments reported in the European scientific journals.

Henry's time had come.

In the year 1826, he learned that the Englishman William Sturgeon had succeeded in making an electromagnet. Henry saw the new form of magnet while on a visit to New York during the period when he was preparing to assume his teaching duties. The purpose of this visit to New York is not certainly known. Were we judging the conduct of an ordinary young man of the time, there would be little reason to seek for an excuse. Henry had a little spare time and some

---

[15] Biographical memoir. *Memorial of Joseph Henry,* p. 441.

loose cash in his pocket for the first time in his life. When these factors are found together, the festive spirit is not far distant. The newly appointed professor may then have gone to New York for no other purpose than to see the sights.

Joseph Henry was not an ordinary young man. His idea of a festivity was more likely to take the form of visits to Eaton's friends, and to call upon Dr. Renwick of Columbia College. In this case, Renwick would certainly have told him that James Freeman Dana, who had moved from Dartmouth and was lecturing at the College of Physicians and Surgeons (then a brick building on the north side of Barclay Street, near Broadway), had one of the new electromagnets.

Henry was fascinated with Sturgeon's magnet when it was shown him. Here were dry leaves for imagination to light and genius to fan into flames. As soon as he returned to Albany he constructed one. No especial difficulties arose. Sturgeon's magnet was so simple one wonders why no one had thought of the idea earlier. A bar of soft iron was bent into the shape of a horseshoe, varnished for insulation, and a copper wire was loosely wound around the bar. The ends of the wire were dipped into cups of mercury connected to the plates of a Voltaic battery. When the circuit was closed, the magnet was capable of sustaining a weight of nine pounds.

Simple as was the construction, let us not overlook the difficulties which Henry had to confront in making his early magnets. Albany may have been a useful town because of the ideas its intelligent population were willing to discuss, but the small town of twenty-four thousand souls suffered from the handicaps of a frontier village when it became a matter of obtaining materials. There were occasions when Henry had to postpone his experiments, as when a diligent search failed to uncover a scrap of zinc in the neighborhood.

Nor should we forget Henry's very modest financial means. An observant reader will share the author's wonder that he could acquire all the copper wire which figures in his experiments. This matter of expense cannot be dismissed lightly. It moulded to some extent the nature of his work. In the first account of his investigation into electromagnetism he begins by deploring that popular lecturers have neglected the subject, and goes on to say: "A principal cause for this inattention to a subject offering so much to instruct and to amuse is the difficulty and expense which formerly attended the experiments—a large galvanic battery, with instruments of very delicate workmanship being thought indispensable." [16]

There was justification for the assumption that considerable expense was involved when it is recalled that Davy's enormous batteries and one built for Hare which comprised 700 plates (measuring 4″ x 3″) were fully described in the journals. A man of modest means was deterred from experimenting because of the expense to which he would be put in duplicating this elaborate apparatus. Henry never had much money to spend on apparatus. No one appears to have devoted much attention to this aspect of Henry's work, but we shall have occasion to observe that it figures strongly in almost everything he did.

He had still another reason for not wishing to possess elaborate equipment. He had no place to put it. The Academy had no laboratories for its professors and students. Henry was assigned the southwestern room on the third floor for his classes. A long table served for demonstrations. Facing this were rows of wooden benches for the students. The room contained no facilities for laboratory work nor any provision for storing apparatus.

Furthermore, the room was in constant use. Some classes began as early as six o'clock in the morning and others were conducted for the benefit of evening students. In between was very little time for original research into any field of science. The only time the young professor could look forward to for a period of uninterrupted personal experimentation was during the month of August, when the pupils were released for vacation. Then, he was granted use of the lecture hall.

The summer of 1826 had been fully occupied, and no time was left for the construction of an electromagnet. When the Albany Institute reconvened for its next season, Henry did not address the members on magnetic properties. The only paper he read that year was on the subject of "Flame," drawing principally upon material obtained while assisting Dr. Beck in his chemical experiments.

Joseph Henry's original scientific work began in earnest when, in August of 1827, the students were all dismissed for a month. He had been able to read about Sturgeon's magnet and some experiments

---
[16] *S.W.* Vol. I, p. 3.

performed with it, in the *Annals of Philosophy*.[17] It was his practice to repeat for his own enlightenment the experiments that others had performed. Before the August vacation arrived, he could see how Sturgeon's magnet could be improved.

At the meeting of the Albany Institute on October 10, 1827, he was able to present his audience with demonstrations surpassing anything of the kind given elsewhere. His paper was called "Some Modifications of the Electro-Magnetic Apparatus," [18] and it shows him in an entirely new role, that of the original experimentalist.

The paper opens with the words: "The subject of electro-magnetism, although one of the most interesting branches of human knowledge and presenting at this time the most fruitful field for discovery, is perhaps less generally understood in this country than almost any other department of natural science." It was a keen mind that saw in electromagnetism, in its state of infancy, "the most fruitful field of discovery." Henry had selected his major scientific field.

Emphasis was laid upon efforts "to form a set of instruments on a large scale which will illustrate all the facts belonging to the science with the least expense of galvanism." This man who had known the pinch of poverty might be expected to take delight in reducing the expense of his apparatus, and the good teacher can be perceived behind his efforts to construct apparatus of a most impressive nature. In this first paper on electromagnetism Henry displayed two qualities which were to characterize his work. He wanted to impress his pupils with models they could easily remember, and the motive of economy is apparent.

No new principles were introduced in this paper, his chief concern being to furnish with a small battery better examples of Sturgeon's experiments. To illustrate Ampère's delicate experiment showing the directive action of the earth's magnetism on a galvanic current when the conductor is free to move (usually illustrated by a small frame of wire with the extremities dipping into cups of mercury), or its modification known as De la Rive's ring (usually an inch or two in diameter), Henry employed much more striking apparatus.

He made a rectangular coil, about six inches by nine, suspended at the end of a silk thread. This coil was wrapped in silk for insulation,

---
[17] New Series. 1826. Vol. XII, p. 375.
[18] Albany Inst. *Trans.* Vol. I, p. 22; *S.W.* Vol. I, p. 3.

and the whole bound together with ribbon. The ends of the wire were soldered to a pair of small plates which dipped into a glass of dilute acid. When the plates were lowered into the acid, the coil, after a few oscillations, settled at right angles to the earth's magnetic meridian.

In a subsequent paper [19] he described how he continued to improve his apparatus. While other men were still using a single conducting wire he obtained better results with coils, and his experiments were seen by large audiences. His De la Rive ring grew to be a coil of 60 feet of wire, covered with silk, and coiled into a ring 20 inches in diameter.

Next he concerned himself with an increase in magnetic force. He took a small piece of soft iron, ¼ inch in diameter, bent into the shape of a horseshoe, and coiled 35 feet of insulated copper wire tightly around this in about four hundred turns. The ends of the wire were soldered to two small plates. This magnet was found to be much stronger than one made up with the customary loosely coiled wire attached to 28 plates, each eight inches square. This magnet was introduced to the Albany Institute in 1829.

Sturgeon's arrangement did not take advantage of Ampère's deduction that electrically produced magnetism passed around the magnet like rings placed at right angles to the surface. Since the wire had to be turned about the insulated core in such manner that there was no metallic contact, the coils were placed at an angle of approximately 30 degrees to the axis of the core, so that each tended to produce its own magnetic field inclined to the axis. Henry's idea of insulating the wire instead of the core permitted him to wind more turns and to place each one at an angle of 90 degrees to the core.

The first stride toward original accomplishment had been taken.

August of 1828 produced the little magnet which was the forerunner of the spool or bobbin solenoid which was to be universally employed. It is doubtful whether he devoted the entire month of his vacation to this development, for this was the year when Eaton had said a party consisting of himself, Lewis Beck, George Clinton, and Henry were proposing to pass through the Erie Canal. The letter containing this proposal was dated February 26th, so that it admitted a wide margin of time for a change in plans. The excursion may have been abandoned. It is mentioned here largely because of the uncer-

[19] *Amer. Jour. of Sci.* Jan. 1831. Vol. XIX, p. 400.

tainty which attends all of Henry's actions and work at this early stage of his career.

Unfortunately for historians, Henry kept no account of the manner in which he passed the vacation months in 1828, 1829, and 1830, for much of his claim to the priority of his greatest discovery depends upon the experiments he performed in the two latter years. Instead, we have nothing but a patchwork of hints. But though he published no more scientific papers, his work was claiming notice. After the demonstration of his improvement on Sturgeon's magnet, Union College conferred upon him the degree of Doctor of Laws. The authorities of the Albany Academy had also awakened to recognition of his abilities and had provided him with better facilities for work. The additional space made available to him was modest enough in all conscience, but it furnished Henry with the nearest approach to a laboratory he had yet obtained. It was a basement room, with a heating stove so that he might work there on winter nights.

He also received an assistant in the person of a former student, George W. Carpenter. Carpenter was young and inexperienced, but he willingly aided in preparing the apparatus for Henry's experiments. If only Carpenter had kept a journal of the work which he performed, he could have clarified some of the disputed points relating to the times at which Henry made certain discoveries, but he was already an old man when he was consulted on the critical dates and his recollections were neither vivid nor accurate.

The year 1830 was a busy one for Henry. The industrious young man nevertheless found time for society and romance, as well as for science. Life in Albany may have had drawbacks for a young scientific scholar, but it had some compensations. The little town had a spirit and a social life all its own. This small community was proud of the young professor who had shown ability and enterprise, credit for which would reflect upon the town and its citizens. In a modest way he had become something of a public figure.

One home in Albany had come to have a special attraction for him. His mother's brother, Alexander Alexander, after conducting a successful business in Schenectady had recently died. His widow and family came to live in Albany. Stephen Alexander, Joseph's cousin, was a delicate and sensitive youth who, at the age of eighteen had graduated from Union College with high honors in mathematics and astronomy. He was to be a teacher at the Academy for a few years. Since they had kindred interests, a close intimacy sprang up between the two cousins. Stephen was to follow Henry to Princeton, and eventually he became a professor of astronomy there. The cousins conducted several investigations together.

But Stephen was not the sole cause of attraction that drew Joseph to his aunt's home. The chief pleasure he derived from his relatives' migration to Albany radiated from the presence of his cousin Harriet. In the midst of his work, Joseph had contrived to find time to visit Schenectady in order to pay court to his cousin.

Their courtship as revealed in a few letters would not seem to have differed from that of any other young couple of their times. There were the same joys and the same misunderstandings. There is barely a pulse of life discernible beneath the formalism, dignity, and respectability of their correspondence, with its shy graces and seemly reticences. Only one letter from Harriet Alexander has been found and from it we gain the impression that the young people had arrived at some views upon a union before the Alexander family moved to Albany.

Schenectady,
February 11, 1828.

My Dear Joseph:

I received yours on Saturday last and would have answered it immediately had not a sore finger disabled me.

It was quite an agreeable although rather unexpected favor, and met with a cordial reception as a proof of your continued esteem and a relief from many unpleasant emotions related to the impression my, at best, foolish and inconsistent conduct may have made on you at your last visit, particularly as your grave phiz and cold air at parting bespoke no little dissatisfaction, and which I will acknowledge, for a moment heightened my displeasure, but on very little reflection, soon convinced me that I was the offender.

I can scarcely tell what at first displeased me. Some mere trifle which ought to have passed unheeded. Impute it not to any act of yours but to the inability of a temper which is often a difficulty. I regret the injury which I fear it has occasioned you, but I fear I should be half inclined to repeat the offence if I could be assured of its being followed by such a pleasing communication.

Mary Ann La Grange is with us at present, to whom I have made some important disclosures but strange to tell she evinced no surprise.

Harriet.

Joseph's letters are written in the same sober tone. His manners are so impeccably correct that they succeed in suffocating his inner feelings. It is apparent that he bent his thoughts to love as sternly as he inclined them to his scientific pursuits. The only letter of his worth quoting was written on the Christmas before their marriage.

Albany,
Dec. 25, 1829.

My Dear Harriet:

For the last three weeks I have been constantly looking forward to this day with the liveliest anticipation of hope, and in truth no Christmas since the days of my boyhood has ever appeared so pleasant in prospect as this.

But now that it has come instead of hastening to enjoy it with you as I fondly anticipated, I am unexpectedly obliged to embark in a few minutes for New York. My stay in that city, however, will be no longer than is necessary to transact some business for the Academy, and if nothing happens I shall probably be in your presence next Wednesday evening, when I hope soon to lose in the pleasures of your company the recollection of every circumstance but the one in which I base my every hope of future happiness.

Do me the favor to accept the accompanying annual and be assured that with sentiments of the highest esteem and feelings of the warmest kind, I am my dear Harriet

Yours devotedly
Jos. Henry.

Joseph Henry was never a self-assertive creature, but Harriet was so self-effacing that their marriage caused no interference with the intensive research program he had planned for that year. They went to live after marriage at 105 Columbia Street, a modest dwelling within sight of the Capitol building. This house made a gallant struggle for survival but, such is the irony of human affairs, was torn down just before Albany proudly celebrated the Henry centennial.

The union of the two cousins was an unusually happy marriage. If Henry's personal life is little known, and we have to be content with only fleeting glimpses of the man, the existence of his family is wraithlike. They exist only in his reflection. None the less, Harriet Alexander Henry seems to have been the ideal companion for Joseph. There were six children born of their marriage; two died in infancy, and the only son, William Alexander, died in 1862 on the threshold of a promising manhood. The declining years of Joseph and Harriet were enriched by the devotion and affection of the three surviving daughters, Helen, Mary, and Caroline.

Modest, retiring, and understanding, Harriet furnished a domestic atmosphere into which Henry could retire in an environment of complete sympathy, forgetting the outer world and its pressures. Not once does family life intrude upon Henry's career.

Henry, like Newton, needed a spur to provoke him into publishing the results of his researches. It was characteristic of him not to publish his results as soon as he obtained them, but rather to wait until he had tested and investigated the various ideas which came into his mind on any one subject, so that he might be convinced he had exhausted that line of thought. It seems reasonable to suppose he saw so much ahead of him by the end of August 1830, especially the construction of new apparatus, that he judged it best not to publish what he thought must be only the preliminary stages of his studies. It was the publication of an account of a rival's achievement which spurred him to reveal how much further he had gone.

Gerard Moll of Utrecht University had been shown Sturgeon's magnet while on a visit to King's College in London. Like other physicists who saw the appliance, he was impressed by its potentialities. Vainly searching the scientific journals for any record of modifications or amplifications of Sturgeon's invention, he decided to construct a magnet and to conduct some experiments of his own devising. He was under the innocent delusion he was improving upon the only form of electromagnet in existence.

By greatly increasing the number of turns in his wire, and by enlarging the area of his battery plates to increase excitation, he was able to make a magnet capable of sustaining weight of 75 pounds. Encouraged by this initial effort, the Dutch professor persevered with

WORK AT ALBANY

his idea and eventually constructed a magnet with a lifting power of 154 pounds.

Moll had benefited by winding his coil of wire more closely, but he did not advance beyond the employment of a single layer of wire. He induced a strong magnetic force by using a powerful battery. The plates of Sturgeon's original battery had an active surface of 130 square inches, and those of Moll were increased to 170 square feet.

When he had the strongest electromagnet which Europe had produced, Moll announced the result of his labors in Brewer's *Edinburgh Journal of Science*.[20]

As soon as Henry was notified of Moll's accomplishment he was prodded into preparing a paper which betrays all the traces of haste in its preparation, and he dispatched it for publication in Silliman's *Journal*.[21] Since his own performance showed a marked advance in comparison with that of Moll, he enclosed a copy of the latter's article, which was published alongside his own.

This paper, entitled "On the Application of the Principle of the Galvanic Multiplier to Electro-Magnetic Apparatus, and also to the development of great Magnetic Power in soft Iron, with a small Galvanic Element," is one of Henry's striking contributions to scientific literature. It is worthy of examination in detail.

Before embarking upon the account of his latest experiments, and because he was addressing a wider and more critical audience than the Albany Institute afforded him, Henry is at pains to convey the idea that he has been working in the field for some time and has made progressive improvements in the electromagnet. He briefly recapitulates his earlier accomplishment. The modification of Schweigger's "multiplier" is described: "The coil is formed by covering copper wire, from $\frac{1}{10}$ to $\frac{1}{20}$ of an inch in diameter, with silk. . . . The effect is multiplied by introducing a coil of this wire, closely turned upon itself."

This was the method employed by Henry in constructing his first electromagnet in 1827, using the coil with an air core. He pointed out that "this coil was applicable in every case where strong magnets cannot be used."

The later experiments were conducted with the object of increasing "the size of electro-magnetic apparatus, and to diminish the

[20] New Series. 1830. Vol. III, p. 214.
[21] *Amer. Jour. of Sci.* Jan. 1831. Vol. XIX, p. 400; *S.W.* Vol. I, p. 37.

necessary galvanic power. The most interesting of these, was an application to a development of magnetism in soft iron, much more extensively, than to my knowledge had been previously effected by a small galvanic element."

Although this is not the sole object of the experiments described in the paper, toward the close he reaffirmed it. "The principal object in these experiments was to produce the greatest magnetic force with the smallest quantity of galvanism."

This affirmation was to assume an importance altogether beyond its merits when Morse claimed that he alone was the inventor of the electromagnetic telegraph, and pointed to this statement of Henry's to disparage the assertion that Henry had been employed in any investigation of the principles of the telegraph.

The picture of his work as presented in this paper is not at all clear. Henry's thoughts upon the phenomenon he was attempting to describe had not crystallized. He had opened up an entirely new field. Moreover, he had no units of measurement other than those of mechanics. No electric units, either of current or potential, had been thought of in connection with galvanism. There was not even a clear idea of electrical resistance. Ohm's law was not yet known to the English speaking world; Henry did not learn of it for another seven years.

Therefore, it is not surprising that Henry's work was qualitative. He observed different kinds of effects, and he used numerical values, such as weights and lengths, to supply a means of comparison. There is no use of mathematics.

Secondly, Henry had not intended publication of his results until he had clarified and classified his views. The publication of Moll's paper had prodded him into revealing his work, and this haste may account for the vagueness of some of the ideas in his paper. However, Henry's report on his experiments is forceful and accurate, so that anyone who wished to repeat his experiments could do so with ease.

Had Henry had the time to tabulate his results he would have had a much clearer view of what he had accomplished, and he could have given his readers a better conception of the progress of electromagnetism under his impulse. A series of tables have been compiled to facilitate our readers' grasp of the facts which were presented by Henry, and these are presented in an appendix.

It should be noted that the battery plates employed were without

WORK AT ALBANY

exception made of copper and zinc. Plain copper and zinc batteries were subject to marked polarization and to variation of electromotive force. In most cases the electrodes (although this word had not yet been introduced) were dipped in a solution of dilute acid, which would give a voltage of about 1.07 volts, in modern terminology. The area of the plates was varied, and internal resistances depended upon the area employed. In one experiment, No. 7, Henry used a Cruickshank trough containing 25 pairs of plates. We may now turn to the experiments themselves.

In one group of experiments Henry used a soft iron bar, ¼ inch in diameter, bent into horseshoe shape. He neglects to give the length of the bar but it is assumed to have been 9 inches, for this is the length given of a bar used in a related experiment. The assumption of this length is consistent with the results obtained. The weight of the bar, then, would be approximately 2 ounces. This bar was wound with 35 feet of copper wire, making about 400 turns. It is difficult today to imagine that these trifling affairs were the first multi-turn multi-layer iron-core magnets ever made.

For other experiments in this group, Henry used an iron bar 10 inches long and ½ inch in diameter.

The tabulated results of this group of experiments will be found in Table 1 of the Appendix.

Henry's object was to create the greatest magnetic power with the least expense of galvanism. He therefore made tests with various electrode areas. Not unnaturally it was observed that, as the battery was made larger the magnetic lifting power was increased. It was, however, noted that a magnet constructed with many turns of silk-insulated wire was "much more powerfully magnetic than another of the same size."

Here we touch upon an important discovery associated with this first group of experiments. A few words must be said upon the dawn of the idea in his mind, since the tale is mildly reminiscent of the circumstances which gave rise to Archimedes famous exclamation of "Eureka!"

Various stories of the episode have been given but we prefer that told by one of his favorite pupils, Dr. H. C. Cameron, professor of Greek at Princeton.[22] He tells us that Henry was seated with friends

[22] Reminiscences of Joseph Henry. In the *Memorial*, p. 173.

in his home. The conversation had momentarily languished, and Henry fell into a deep reverie. Suddenly, he startled his companions by bringing his hand down heavily on the table, exclaiming, "I have it!"

Since these illuminative tales of famous men are usually apocryphal, and this one exhibits Henry so unlike his customary calm unruffled self, a friend once questioned him upon the authenticity of the incident. Somewhat bashfully he confessed, "When this conception came into my brain I was so pleased with it that I could not help rising to my feet and giving it my hearty approbation." [23]

The idea which had suddenly flashed into his mind is thus described in this paper: ". . . a very material improvement in the formation of the coil suggested itself to me. . . . It consisted in using several strands of wire, each covered with silk, instead of one:—agreeably to this construction, a second wire, of the same length as the first, was wound over it, and the ends soldered to the zinc and copper in such a manner that the galvanic current might circulate in the same direction in both, or in other words that the two wires might act as one."

He said the idea occurred to him "on reading a more detailed account of Professor Schweigger's galvanometer." Schweigger never suggested using multi-layers of wire. The idea might have been derived from a totally different source in which a galvanometer constructed by Oersted was described. This will be found in Jacob Green's *Electro-Magnetism, etc.,* published in 1827, in which the galvanometer is described as having been "wrapped or wound throughout its whole length in silk thread. Thus all electric communication is avoided between the different parts of the wire, the turns of which are wound one over the other." [24]

Of course, Henry may never have seen this particular detailed description but, in any event, insulated wire had not been applied to magnets. They had been constructed by insulating the core, and winding on it a single spiral of uninsulated wire. In his earlier work, Henry had followed Sturgeon's practice of winding a single wire on the insulated core. It is not clear whether he ever wound coils separately with the intention of applying them to magnet cores after winding. This form of construction was used by Robert Hare as

[23] Mayer, A. M. Henry as a Discoverer. *Memorial*, p. 483.
[24] P. 81.

early as 1831, in making the Henry type of magnets. In a letter dated March 4, 1831 he describes in detail the process of winding the coil on a mandril and of later slipping it over the core.[25] Apparently he had in mind an improvement in construction detail.

Ames gives an interesting account of the manner in which Henry obtained his insulating materials:

> I have heard from the family that the silk used by Henry was really a series of ribbons of silk obtained by the sacrifice on the part of his wife of her white silk petticoat. An electromagnet made by Henry and used in his experiments is still in existence, being treasured in the museum at Princeton University, and one can still see the white silk ribbons used in its construction, so I believe the story I have heard is true.[26]

Henry used the same small magnet which had lifted 14 pounds with a single winding of 400 turns, but wound upon it another 400 turns. This was connected in parallel with the first. The double-layer magnet lifted 28 pounds.

This was Henry's first "quantity" magnet, and one of his most notable developments. By amplifying the multi-turn multi-layer coil and the "quantity" connection, he was soon to astonish the scientific world.

In summarizing the first group of experiments, Henry arrives at an accurate explanation, considering that he had no electrical measuring units, and that he regarded electricity as "an imponderable fluid." He says:

> The experiments conclusively proved that a great development of magnetism could be effected by a very small galvanic element, and also that the power of the coil was materially increased by multiplying the number of wires, without increasing the length of each.
>
> The multiplication of the wires increases the power in two ways; first by conducting a greater quantity of galvanism, and secondly by giving it a more proper direction, for since the action

---

[25] *Amer. Jour. of Sci.* July, 1831.
[26] Certain aspects of Henry's experiments on Electromagnetic Induction. *Science.* Vol. 75, 1932, p. 89.

of a galvanic current is directly at right angles to the axis of a magnetic needle, by using several shorter wires, we can wind one on each length of the bar to be magnetized, so that the magnetism of each inch will be developed by a separate wire; in this way the action of each particular coil becomes very nearly at right angles to the axis of the bar, and consequently, the effect is the greatest possible. The principle is of much greater importance when large bars are used. The advantage of a greater conducting power from using several wires might, in less degree, be obtained by substituting for them one large wire of equal sectional area, but in this case the obliquity of the spiral would be much greater and consequently the magnetic action less; besides this, the effect appears to depend in some degree on the number of turns which is much increased by using a number of small wires.

Several small wires conduct more common electricity from the machine than one large wire of equal sectional area; the same is probably the case, though in a less degree, in galvanism.

It will be seen that this group of experiments marks a quite considerable advance in the knowledge of electromagnetic phenomena. Henry had noted how the addition of the second coil exactly doubled the lifting power of the magnet, but he did not say that the lifting power was directly proportional to the number of turns or to the number of coils. He said that "the effect appears to depend in some degree on the number of turns," inferring this was the answer. It is to be regretted that he did not think mathematically, or that he was not inspired to do so by the perception of this relationship. Had he investigated the results mathematically he would have uncovered much more about magnets and the magnetic properties of iron. However, following the path of the qualitative experimenter he did investigate the matter further, as we shall see, in the fourth group of experiments.

The second group of experiments described in this noteworthy paper do not deal directly with magnetism. They have the aspect of being preliminary studies for the third group which is concerned with the development of magnetism at a distance, and contains a discovery of the greatest consequence relating to the electromagnetic telegraph.

In this second group Henry had the assistance of Dr. Philip Ten

Eyck, a young man of good family in Albany and of whom we shall have a word to say later. He is mentioned here not only because of his aid in the experimental work, but also as the probable provider of the unusual quantity of copper wire which figures in the work. There is no clue to the source of the supply, but it is not illogical to assume that Ten Eyck helped to finance the purchase, since such an acquisition must have been beyond the means of Joseph Henry.

In this group of experiments Henry employed as a measuring instrument a galvanometer of the Schweigger type, which consisted of eight feet of wire wound into a flattened coil and a magnetic needle suspended on a silk thread, probably suggested by the Hansteen suspension for magnetic needles used in measuring the earth's magnetic intensity. The current from a single pair of copper and zinc galvanic plates, two inches square, was passed through various lengths of wire, and the effect on the galvanometer was noted.

Although the galvanometer had been in use then for ten years, no one knew what it measured. Crudely, it measured "effect," and the degree of effect was expressed in terms of the needle's angle of deflection. Efforts had been made to interpret this angle, and William Ritchie was prominent in these efforts. Henry alludes to Ritchie's work, but the allusion betrays once more the haste with which he prepared this paper, since he not only mentions an erroneous source of information but misinterprets Ritchie's result.

However, the two experiments in this group are not without their significance. Rather more than ⅕ of a mile of copper wire (1060 feet, to be exact), already known as "bell-wire," [27] was stretched several times across the large lecture room of the Academy. Current from the single pair of plates was passed through this wire, and its effect on the galvanometer noted. The mean of several observations showed the deflection of the needle to be 15 degrees.

Then the current from the same pair of plates was passed through half the length of wire (530 feet) and the deflection was noted to be 21 degrees.[28]

---

[27] Because it was used for the bell pull of mechanical bells.

[28] Henry adds a note: "by reference to a trigometrical table it will be seen that the natural tangents of 15° and 21° are very nearly in the ratio of the square roots of 1 and 2, or of the relative lengths of the wires in these two experiments." Tan 21°/tan 15° $= 1.43$; $\sqrt{2} = 1.41$. Henry thought this agreed with Ritchie's result, but Ritchie had said that the deflecting force was proportional to the sine rather than to the tangent of the deflection angle, the sine relation having been pointed out by Antoine Cesar Becquerel five years earlier.

In the light of our current knowledge we might very well ask why Henry did not pursue this investigation further, but again we must point out that he was groping his way among new truths.

He had a definite object: to ascertain how the length of wire was related to the "effect." He learned what he required to know at the moment, and passed on to the main experiments, which are found in the third group. However, having regard to the nature of Henry's mind, one cannot refrain from wondering why he did not pursue the question of the relationship between current and length of wire. It was unexplored territory and its mystery ought to have beckoned him. He did not try a variety of lengths of wire and observe their effect upon the galvanometer. The fact that he did not make these additional experiments stamps him irredeemably as a scientist who was satisfied with qualitative knowledge.

One might add the comment here that in Group 1 of the experiments, designed to observe the effect of different batteries on the lifting power of a magnet, two magnets were used, thus making interpretation difficult. In Group 2 only one variable was employed. In the next group, designed to test the lifting power of magnets at different distances from the battery, he confuses the results by introducing two variables again—distances and battery powers, for he uses three batteries. But in spite of his quantitative deficiencies, Henry contrived to make some remarkable discoveries with these experiments.

In this Third Group (see Table 2 of Appendix), the galvanometer was detached and the 1,060 feet of wire was connected to the small magnet made of ¼-inch iron. This magnet had eight feet of copper wire wound around it. The magnetism in the small horseshoe proved to be barely perceptible. The small plates of eight square inches were removed from the battery, and a piece of zinc plate, 4 × 7 inches surrounded by copper plate of undetermined dimensions was substituted. When this was connected directly to the magnet, the weight lifted was 4½ pounds. When the current was passed through the 1,060 feet of wire it barely lifted ½ ounce. Then the same current was passed through 550 feet of wire. This lifted 2 ounces, and when two wires of the same length were joined in parallel to the battery the magnet lifted 4 ounces.

At this point Henry had an astonishing intuition. He made a radical change in his battery, using a Cruickshanks trough containing 25 double plates, offering the same area of zinc to the acid as the battery used in the last experiment. In this case the weight lifted was 8 ounces. When the long length of wire was removed and the trough attached directly to the magnet, the latter lifted only 7 ounces.

The anomaly in this pair of experiments was somewhat perplexing. Henry doubted the validity of his results. He thought that the smaller power derived from the shorter connections might be attributed to the drying of the trough. He would have liked to pursue this series of experiments had there been time, but the vacation was over. The long wire had to be reeled in and school work resumed. He summed up the results of the experiments in these words: "On a little consideration, the above result does not appear so extraordinary as at first sight, since a current from a trough possesses more "projectile force," [29] to use Professor Hare's expression, and approximates somewhat in intensity to the electricity from the common machine. May it not also be a fact that the galvanic fluid, in order to produce a greater magnetic effect, should move with a smaller velocity, and that in passing through one-fifth of a mile, its velocity is so retarded as to produce a greater magnetic action? But be this as it may, the fact that the magnetic action of a current from a trough is, *at least,* not sensibly diminished by passing through a long wire, is directly applicable to Mr. Barlow's project of forming an electro-magnetic telegraph, and it is also of material consequence in the construction of the galvanic coil."

Before examining the importance of the deductions derived from these experiments, let us pause to comment upon that allusion to "Mr. Barlow's project." Peter Barlow, who taught mathematics at the Woolwich Military Academy in England, was the first to test the practicability of Ampère's suggestion that by sending the galvanic current through long wires connecting distant stations the deflections of enclosed magnetic needles would constitute very simple and efficient signals for an instantaneous telegraph. Barlow published the results of his test in the *Edinburgh Philosophical Journal* and so far from furnishing encouragement toward the realization of Ampère's suggestion, he was very emphatic in his declaration against it. He

---

[29] That is to say, more electromotive force. Note that Henry was thinking in terms of motion of a fluid according to Newton's laws.

announced: "I have found such a sensible diminution with only 200 feet of wire, as to at once convince me of the impracticability of the scheme." [30]

Henry's unfortunate allusion to "Mr. Barlow's project" was to be seized upon by Morse as proof that Henry was not well informed upon the subject of the electromagnetic telegraph.

The experiments in this third group are important, for the knowledge gained from them points conclusively to a matter of material consequence in the construction of the coil. He says: "From these experiments it is evident that in forming the coil we may either use one very long wire or several shorter ones as the circumstances may require: in the first case, our galvanic combinations must consist of a number of plates so as to give 'projectile force'; in the second, it must be formed of a single pair."

This is the first allusion that had been made to accommodating the battery connection to the circuit. Henry is here forming the conception that was to be one of his greatest contributions to electrical science, the idea of associating "intensity" and "quantity" coils with "intensity" and "quantity" [31] circuits. He had not yet given the connections the names by which they were to be known throughout the next generation, but the allusion is too significant to be passed in silence. He was very close to a conception of Ohm's law, but his qualitative descriptive approach kept him from expressing his notions in numerical ratios.[32]

Comment upon the third group of experiments must center upon Henry's quickness to grasp the general relationships between the essential factors. But because his thinking was descriptive and in

[30] *Edinburgh Philosophical Journal,* Vol. XII, p. 110, Jan. 1825. There is a good deal of confusion about Barlow's exact interpretation of his own experiments, but there is no doubt that his statement of failure discouraged others from working on the telegraph. In his discussion of Barlow's "law of diminution" W. B. Taylor (*Memorial of Joseph Henry,* footnote, p. 224) says that Barlow "was led (erroneously) to regard the resistance of the conducting wire as increasing in the ratio of *the square root* of its length." If this is so, Barlow should have been encouraged, for this would make the resistance even less than directly proportional to length, which is the actual case. Perhaps Barlow meant the *square* of the length, which would be discouraging indeed.

[31] "Intensity" and "quantity" circuits are now known as "series" and "parallel" respectively.

[32] Ohm's original statement of his "law" was published in pamphlet form in Berlin in 1827. As late as 1835, Henry asked his friend Bache for information upon Ohm and the law associated with his name. Bache's reply was depressing. He did not know of Mr. Ohm, "he was probably one of those outlandish men who wrote in the unknown tongue—German."

terms of mechanical analogies, he misses a significant feature which his results would have revealed through mathematical analysis. Had he plotted the weights lifted against the length of connections for some of the experiments, he would have obtained a suggestive curve. Or had he plotted the logarithms of the weights against the lengths, he would have had reason to suspect an exponential relationship which would have justified further investigation. They fall nearly in a straight line.

The mathematical tools were available which would have enabled Henry to enlarge his experiments had he chosen to examine the intrinsic and essential rather than the obvious and apparent. But he refused to be diverted. The time had not yet come for mathematical analyses.

The fourth group of experiments was an extension of the first group upon a large scale. Henry's object was to "produce the greatest magnetic force with the smallest quantity of galvanism." In the pursuit of this end he met with such conspicuous success that the name of Joseph Henry was made famous in the scientific world. No new phenomena were observed, but a powerful magnet was the outcome, the precursor of the giant magnets of today.

A large magnet was designed especially to test the results of the first group of experiments. A bar of soft iron, two inches square and 20 inches long, was bent into the conventional U form, or horseshoe, to stand 9½ inches high. The sharp edges were rounded with a hammer. This bar weighed 21 pounds.

Then, an "armature or lifter" weighing seven pounds was cut of the same material. This was filed flat on one side so that it would provide a perfect fit against the ends of the horseshoe. It was intended to hang weights on this armature to measure the magnetic force developed.

Around the horseshoe 500 feet of copper bell-wire were wound, not in one continuous coil but in nine separate coils, each of 60 feet of wire. Each coil covered two inches of the iron bar, and the wire was wound back and forth on itself in several layers. The ends of each coil were left protruding and were numbered so that experiments could be made with various combinations of coils by joining the projecting ends.

The magnet was then suspended from a wooden frame, 3 feet 9 inches high and 20 inches wide.

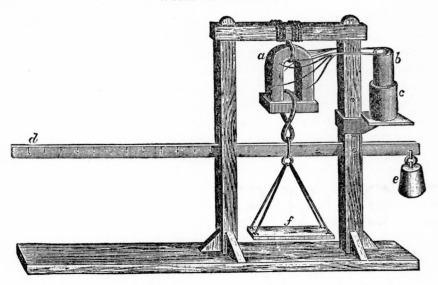

Henry's electromagnetic testing apparatus. "*a*, the magnet covered with linen, the ends of the wires projecting so as to be soldered to the galvanic element *b*. *c*, a cup with dilute acid on a moveable shelf. *d*, a graduated lever. *e*, a counterpoise. *f*, a scale for supporting weights."

A novelty was introduced in the insulation of the copper wire which betokens Henry's desire to strive after economy. The silk ribbon used to wrap the coils had become a matter of considerable expense when long wires were used. At the suggestion of Dr. Lewis Beck (brother of the Academy's principal) it was decided to wrap the coil with linen thread. Beck had also attempted to avoid the cost of the copper wire, and to save the labor of winding it by using iron wire, which was available commercially ready wound. He found, however, that the resistance was too high. His suggestion to wrap the copper wire with linen thread was gratefully adopted and found to answer requirements.

The battery used to excite this new magnet consisted of two concentric copper cylinders with one of zinc between them. The active area on both sides of the zinc was about four-tenths of a square foot. The cylinders were immersed in half a pint of acid.

Henry's first observation was that the lifting power of the magnet was increased as he brought additional coils into action. With any one coil, the magnet supported only 7 pounds, but when two (one coil on each side of the bent bar) were united, the magnetic force would support 145 pounds. The maximum power exerted by two selected coils was 200 pounds, but only the end coils were capable of developing this power.

He now turned his attention to the power source. When all the nine coils were connected, as we would say in parallel, the magnet supported the astonishing weight of 650 pounds. Then the size of the battery was increased, a zinc plate 12 inches long and 6 inches wide surrounded by copper was introduced. This increased the lifting power to 750 pounds.

At this stage Henry had discovered the magnetic saturation of iron, although it is a question whether he should be credited with it, for he observed it without, apparently, attaching any importance to it. He did not entirely overlook this question of saturation, however, for he wrote: "This is probably the maximum of magnetic power which can be developed in this horseshoe, as with a large calorimeter, containing 28 plates of copper and zinc, each 8 ins. square, the effect was not increased, and indeed we could not succeed in making it lift as much as with the small battery."

Henry summarizes the results of this group of experiments [33] as follows:

> The strongest magnet of which we have any account is that in the possession of Mr. Peale, of Philadelphia:[34] this weighs 52 pounds and lifted 310 pounds, or about six times its own weight. Our magnet weighs 21 pounds, and consequently lifts more than thirty-five times its own weight; it is probably, therefore, the most powerful magnet ever constructed.
>
> This, however, is by no means the maximum, which can be produced by a small galvanic element, as in every experiment we have made the power increases by increasing the quantity of iron; with a bar similar to the one used in these experiments, but of double the diameter, or of eight times the weight, the power would doubtless be quadrupled, and that too without increasing the size of the galvanic element.[35]

---

[33] Tabulated in Table III of the Appendix.

[34] Franklin and Titian Peale were proprietors of the Museum of Art and Science, founded by their father, Charles Peale, the well known artist, who had died in 1827.

[35] For some reason, probably haste, Henry did not express this accurately. In order to get eight times the weight the length as well as the diameter would have to be

WORK AT ALBANY

When Henry set out to produce "the greatest magnetic force with the least expense of galvanism" he can scarcely have believed that he would immediately arrive at the construction of the most powerful magnet in existence at the time. We would probably attribute more to him than is justifiable in supposing that he anticipated such a result, for it must be remembered that the production of this magnet capable of sustaining 750 pounds, especially with such a diminutive battery, was a prodigious accomplishment. He had converted Sturgeon's "philosophic toy" into an instrument of infinite possibilities. It raised him at one bound to the forefront of American scientists.

The achievement brought instant recognition, not only among the scientists, but also among a few alert industrialists who perceived how a powerful magnet might be put to practical use. As Henry's story unfolds it will be seen that his large magnet and other revelations in this paper unleash several forces. But we have not yet exhausted the resources of this remarkable paper. Some other experiments must be examined.

Experiments were performed to examine the same phenomena with modified apparatus. They lack the dramatic qualities of those described, but they must not be overlooked since they illustrate the thoroughness of Henry's methods.

The iron bar was rewound with six coils, each 30 feet long. When these were connected in parallel, the lifting power of the magnet was 375 pounds. With the same coils connected in such a way that he had three coils, each of 60 feet in length, and using the same battery, the magnet would only sustain 290 pounds. He was not surprised at the result, and was satisfied to make the remark that "from this it appears that six short wires are more powerful than three of double length."

What did surprise him, however, was the weakness of the attractive force of a single pole of his magnet. "In these experiments a fact was observed, which appears somewhat surprising: when the large battery was attached and the armature touching both poles of the magnet, it was capable of supporting more than 700 pounds, but when only one pole is in contact it did not support more than 5 or 6 pounds, and in this case we never succeeded in making it lift the armature

---

doubled. If the diameter only were doubled, the weight would be quadrupled; probably Henry was thinking of this case, so that if the force were proportional to the weight, it would be quadrupled also.

(weighing 7 pounds). This fact may be common to all large magnets, but we have never seen the circumstances noticed of so great a difference between a single pole and both."

Thus he came very close to perceiving a magnetic circuit, but the conception completely eluded him. He had so much to think about in the midst of all these observations that it is no wonder some rather important features escaped his attention.

During the course of this work Henry had the companionship of Dr. Philip Ten Eyck,[36] who was possibly now his most intimate friend as well as his most intelligent collaborator. He helped Henry by sharing much of the labor in winding coils and making small parts of the equipment during the hours when Henry was engaged with his classes. When Henry vacated his position at Albany, Ten Eyck was appointed his successor. The name of this young doctor must be mentioned here, for he must have contributed from his purse in these experiments as well as having played an active part in them.

Under Henry's supervision Philip Ten Eyck performed a series of four experiments which are described. In one of these a tiny magnet, one inch long and 0.06 inch in diameter, was made to support 420-times its own weight. "In this last result the ratio of the weight lifted, to the weight of the magnet is much greater than any we have ever seen noticed; the strongest magnet we can find is one worn by Sir Isaac Newton in a ring, weighing three grains; it is said to have taken up 746 grains or nearly 250 times its own weight. M. Cavallo has seen one of 6 or 7 grains weight which was capable of lifting 300 grains, or about 50 times its own weight. From these experiments it is evident, that a much greater degree of magnetism can be developed in soft iron by a galvanic current, than in steel by the ordinary method of touching."

Altogether, this paper Henry addressed to his first wide audience showed a remarkable progress in electromagnetism. The author reported contributions in two distinct ways. First, he had made considerable extension to previous experiments and, second, he had made a valuable contribution of an original idea. In building a coil of many turns and layers of insulated wire around an iron core, he had enlarged Sturgeon's conceptions of the magnet. But in building an electromagnet of several distinct coils connected in parallel, he had

---

[36] Son of Judge Egbert Ten Eyck who had sat in the State legislature before he was appointed a judge.

created a new type of circuit which gave to the magnet a greater power than hitherto had been possible. In discovering the relations of "intensity" and "quantity" circuits he was undisputed creator of a conception which made possible the electromagnetic telegraph.

Although the scientific world was momentarily more impressed by the spectacular progress he had made with Sturgeon's idea, Henry saw that the distinction between what he called the "quantity" and "intensity" magnets and the recognition of their essential properties was the greater scientific achievement.

The distinction dwells in the realization that the magnet of several comparatively short coils was more effective when connected to a battery consisting of a single pair, or several pairs connected in parallel, since this gave greater amperage. But the magnet constructed of a single long coil was more effective when connected in series with many pairs, which gave the high voltage necessary to drive a large current through the external resistance. The two battery arrangements were called respectively "quantity" and "intensity" batteries and, consequently, Henry called the appropriate types of magnet by the same names to "avoid circumlocution."

Quantity and intensity were soon recognized to be terms applicable to circuits as well as to components of circuits. Although they were convenient terms, and were employed by scientists for the next thirty years, they did not outlast Henry's life. As applied to circuits they were gradually replaced by the terms "parallel" and "series," which have been used here for the sake of simplicity. It is important to observe that in the idea of connecting a quantity magnet with a quantity battery, or an intensity magnet with an intensity battery, we perceive that Henry had arrived at the notion of matching impedances (resistances, in his case) for maximum power transfer.

The year 1831 had begun auspiciously with the publication of this splendid paper in April, and it was to continue to bear a fruitful crop. One of the most remarkable achievements of the year, although it does not figure in any of his scientific writings, was the construction of the first electromagnetic telegraph. It so impressed his students and the visitors who came to Albany for the purpose of talking to the man who made a huge magnet that Henry later had no difficulty in assembling reliable witnesses to his claim that he had demonstrated the telegraph in this year.

WORK AT ALBANY

In developing the electromagnet, the tool with which Faraday and others were to make their own discoveries, Henry had accomplished something which might have caused him to raise his chin when he walked through the streets of Albany had he been anything else but the most modest of professors. After he had succeeded in passing an electric current through a long wire, a happy idea came to him of the manner in which he could demonstrate its possibilities.

As we have already pointed out, Henry took a special delight in impressing physical principles upon his audiences by resorting to surprising effects. A class of boys might not be impressed by the repetition of the experiments he had conducted during the vacation. He wished them to have a demonstration which would remain in their minds and memories.

When the classes resumed, Henry had reeled up his long wire and had carried it up to the third-floor classroom. Here he strung the wire around the walls and even added to its length, for it is described as having reached a length of almost a mile. At one end the wire was connected to the plates in the Cruickshanks trough, and the other end was connected to the magnet. Close to one pole of the magnet he placed a permanent magnet on a pivot so that it was free to swing. When the circuit was closed and the electromagnet was energized it instantly repulsed the permanent magnet and caused one end to swing sharply against a small office bell.

This was the first electromagnetic telegraph, and it possessed the great advantage of having an audible signal.

Henry recognized the nature of this invention. Many others had sought the means of transmitting a signal over long distances, and it is not improbable that he was aware of their efforts and their failures. Taylor mentions [37] the performance of the Spaniard Salva who, in 1798, had transmitted messages electrically from Madrid to Aranjuez, a distance of 26 miles. Such communication was feasible with the high voltage discharge produced by a frictional machine, but no continuous current could be maintained, and there must have been difficulties over insulation. But, thus far, the best efforts had been directed toward making a magnetized needle swing. Henry was the first to think of and to employ the audible signal.

The original little bell of his telegraph was about two inches in diameter, and is still preserved in the Albany Academy. On December

[37] Scientific Work of Joseph Henry. *Memorial*, p. 224.

17th, 1924, Henry's birthday was celebrated by appropriate addresses broadcast from the radio stations of Schenectady and Troy, on which occasion the relic came once more to life and the sound of the little bell was heard by radio listeners all over the United States.

Why did Henry take no steps to ensure for himself the pecuniary rewards he must have known could be gleaned from such a useful invention? He made no effort, then or later, to protect his rights to the discovery of the principle by applying for a patent.

This was the attitude Franklin had adopted, and it had brought him no little praise, but Henry's similar lack of interest in making money from patents has always been regarded as freakish. His unselfish attitude toward acquiring money has been regarded as an odd quirk in his character. Once, later in his life, when he was asked why he had not patented the telegraph, he answered: "I did not then consider it compatible with the dignity of science to confine benefits which might be derived from it to the exclusive use of any individual." Then, he immediately added the comment: "In this, I was perhaps too fastidious." [38]

[38] Letter to Rev. S. B. Dod. *Memorial,* p. 150.

III

THE DISCOVERY OF INDUCTION

Henry's early research, which established his reputation and for which his name is remembered, was of a kind known today as "basic research." It was intended to extend the sum of human knowledge. Another branch of scientific research, "applied research," is intended to use new knowledge for practical ends. Thus an industrial civilization depends on both basic and applied research; without either there could be little progress. Henry distinguished himself in both fields; but since technology tends to grow outmoded and obsolete, while basic truths last indefinitely, he is remembered for his contributions to basic research. At the time, however, in the expanding American industrial society, the discoveries of Watt, Arkwright, and Fulton aroused much more general interest, because they were of obvious immediate practical value. Nevertheless Henry's work received recognition among scientific men, and it was not until his reputation as a worker in basic science was established that he turned to applied research.

Nearly all workers at basic research are motivated by scientific curiosity and love of truth. This cannot be said of applied research, where there is always the possibility of great monetary reward. It must not be forgotten, however, that applied research like basic research may be pursued out of desire to benefit humanity. Henry's main drive in his electromagnetic investigations was certainly love of truth. Later, when he had accepted the secretaryship of the Smithsonian Institution in spite of offers of higher remuneration elsewhere, he directed and pursued both basic and applied research as a duty to humanity. He never sought royalties or commissions for his scientific work.

Having demonstrated that the idea of the electromagnetic telegraph was feasible, Henry abandoned that field of investigation to examine some other phenomenon he had observed. He omitted to say what this was, but he clearly disregarded the opportunity to make money by developing a practical telegraph so that he might be free to follow another scientific will-o'-the-wisp.

The construction of a magnet of unparalleled power had aroused immediate interest. Upon the recommendation of Amos Eaton, their consultant, the Penfield Iron Works near Crown Point, New York, ordered two magnets to be made.[1] These magnets were to be used for the practical purpose of extracting the iron from pulverized iron ore. They were not large magnets but, besides being the first to be used in commerce, they should be remembered for having fascinated Thomas Davenport, the touzle-headed blacksmith of Brandon.

Davenport left an interesting picture of one of these first industrial magnets as he saw it. He described it artlessly as "an electro-magnet weighing about three pounds, to which were attached two sets of cups consisting of copper, or zinc, cylinders to be used in earthen quart mugs."

> The poles of the magnet were placed on the face of a common blacksmith's anvil, and, as soon as the cups were immersed in the solution contained in the earthen mugs, the magnet adhered sufficiently strong to raise the anvil from the floor, and would continue to raise it until the battery cups were removed from the solution of sulphuric acid and water. This was done by raising the cups out of the mugs, and the conductors were permanently connected together by soldering the battery conductors with the magnet.
>
> Here was one of the wonders of Nature and Providence! A new kind of magnetic power! I inquired if the battery cups remained in the solution and one or both the conducting wires should be cut in two, whether, on reconnecting the conductors with the fingers, or otherwise, as to barely bring them in contact, the power

---

[1] Shortly after the inauguration of the industrial application of the magnet, the name of the little town which encompassed the site of the Penfield Iron Works was changed to Port Henry, as a compliment to the Albany professor.

would not suddenly be renewed, and again suddenly destroyed on separating them. The reply was that the power of the magnet would be suddenly destroyed, for they had not long since broken the wires, which connected the cups with the magnet, and supposing the instrument nearly spoiled, packed it up and sent it to Albany for Professor Henry to repair; and it had just returned. I requested the privilege of cutting the conductors and connecting it together for the purpose of ascertaining the fact, but was refused.[2]

Davenport bought the spare magnet for $75 and, having carefully studied its construction, proceeded to investigate its properties. As the result of his investigations he invented a rotary motor, the first electromagnetic railroad car, and the first electromagnetic piano-player. His views upon the new magnets may be accepted as typical of those of intelligent laymen. It was "a new wonder of Nature and Providence." There was less of wonder and more of joy among the sophisticated.

A magnet of much more impressive dimensions and power was promptly ordered by Benjamin Silliman for Yale University, where a notable school of science was growing up under his guidance. This magnet was constructed from a three-inch octagonal bar of soft iron, 30 inches long, and weighing 59½ pounds. When bent into the conventional horseshoe shape it stood one foot high. It was wound with 26 strands of copper wire, each 31 feet long, and wrapped with cotton thread. After allowance was made for the projecting wire ends, the aggregate length of the coils was 728 feet.

Again one pauses to wonder where Henry obtained his copper wire. It cannot have been a commodity in common demand, for, three years later, when Eaton needed such wire in a hurry, a search of all the hardware stores in Troy produced no more than five pounds of it.

The Yale magnet, with a galvanic element presenting an active surface falling just short of five square feet, was able to support a weight of 2,300 pounds. For the period, this magnet was a revelation of stupendous power. Henry's own weaker magnets had inspired a generous tribute from Sturgeon, who wrote: "Professor Henry has

[2] Davenport, W. R. *Thomas Davenport: the Brandon Blacksmith*. Montpelier, Vt. 1929, p. 57.

been able to produce a magnetic force which completely eclipses every other in the whole annals of magnetism; and no other parallel is to be found since the miraculous suspension of the celebrated oriental impostor in his iron coffin." [3]

The Yale magnet far surpassed those which had aroused Sturgeon's admiration. An account of it was given in a letter accompanying it on delivery, and was published in Silliman's *Journal*,[4] where the editor himself observed in a footnote that betrayed his pride, "It is eight times more powerful than any magnet hitherto known in Europe."

Although European scientists did not pay a great deal of attention to what was going on in America, such an achievement could not fail to attract notice and Michael Faraday was one of the first to imitate Silliman in having a quantity magnet built for his work. In France, too, Henry's performance aroused admiration. Pouillet introduced one of his magnets at the session of the Société Philomatique in Paris on June 23, 1832.[5] One may doubt whether Pouillet regarded himself as fortunate in his experience with the electromagnet. Displaying its properties in a lecture to his students, he rashly used his bare hands to break the exciting current so that he might prove the magnetic force had entirely disappeared. He was almost thrown off his feet by the violence of the shock, an embarrassing experience for a college professor.[6]

The staff at Yale were not slow to put their great magnet to the test. Silliman was away at the time, but his assistant, Charles Upham Shephard, could not control his curiosity. He immediately made an experiment he had been waiting to perform as soon as he could command a strong flow of magnetism. He wanted to learn whether magnetism could decompose water as did galvanic electricity. Needless to say, the experiment failed, as it had done with Fresnel who tried it in 1820.

[3] *Phil. Mag. and Annals*. March, 1832. Vol. XI, p. 199.
[4] April, 1831. Vol. XX, p. 201.
[5] *Nouveau Bulletin des Sciences*. 1832, p. 127. Pouillet also described the construction of Henry's magnets in his *Eléménts de Physique Expérimentale*. 3rd edition, 1837. Vol. I, p. 572.
[6] Note that the phenomenon which caused Pouillet's shock was *self-induction*, the discovery of which was Henry's greatest single contribution to science. Henry's paper announcing self-induction, as we shall see, was published in July 1832, and obviously Pouillet did not know about it at the time of his disconcerting experience. But the idea of self-induction did not occur to Pouillet, as it might have had he pondered the reason for his shock.

The next issue of Silliman's *Journal* [7] carried a letter from Henry in which he announced another discovery. "I have lately succeeded in producing motion in a little machine, by a power which I believe has never before been applied in mechanics,—by magnetic attraction and repulsion."

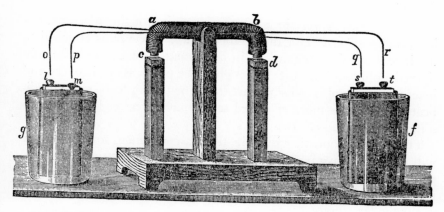

Henry's electromagnetic engine. Henry wrote: "*ab* is the horizontal magnet, about seven inches long, and movable on an axis at the centre: its two extremities when placed in a horizontal line, are about one inch from the north poles of the upright magnets *c* and *d*. *g* and *f* are two large tumblers containing diluted acid, in each of which is immersed a plate of zinc surrounded with copper. *l*, *m*, *s*, *t*, are four brass thimbles soldered to the zinc and copper of the batteries and filled with mercury."

The machine consisted of a bar electromagnet about nine inches long, balanced horizontally so that its poles stood over the poles of two permanent magnets fixed vertically, with like poles upward. Two pairs of terminals were joined to the wire of the electromagnet and bent down so that when the electromagnet was tilted in one sense, one pair of terminals dipped into a mercury cup joined to the copper and zinc of a galvanic cell. The current was then passed through the wire in such a sense that the end of the electromagnet nearest the pole of a permanent magnet was of the same polarity. This end was repelled and raised, so that it lifted the terminals from the mercury cup, while the other end was attracted. The circuit was thus broken but, as the rocking bar moved on, the other pair of terminals was dipped into its other mercury cup, joined to the copper and zinc of

[7] July, 1831. Vol. XX, p. 340.

another galvanic cell. These terminals were so arranged that the current entering them passed through the wire in such a way as to reverse the magnetism, thus bringing about a repulsion and attraction as before, but in the opposite sense. The reciprocating motion thus set up was repeated about 75 times a minute for an hour, or as long as the battery power continued.[8]

Henry did not entertain a very high opinion of this machine, but he went on to say, "Not much importance, however, is attached to the invention, since the article, in its present state, can only be considered a philosophical toy; although in the progress of discovery and invention, it is not impossible that the same principle, or some modification of it on a more extended scale, may hereafter be applied to some useful purpose."

There was prophetic vision in that last sentence, for the reader will doubtless have perceived that the oscillating electromagnet corresponds to the armature of a modern direct-current motor, and in this rocking bar we have the precursor of the rotating commutator of the modern motor.

There was challenge, too, in his final statement. He was virtually suggesting, "Here is something new and useful. Apply a little ingenuity to the idea and it will conduct you to a profitable outcome." When he was constructing this "philosophic toy" Henry was yielding to the prevailing passion for reciprocating machinery which Newcomen had released upon the world and Watt had amplified. But he was too busy with other things to carry his idea to a practical conclusion. He placed a modest valuation on the conception and left its development to others.

The question is sometimes raised whether Henry had not been anticipated by a similar motor made by Salvatore dal Negro of Padua. There is a very brief description of Dal Negro's motor in a memoir entitled "New Electro-Magnetic Experiments and Observations" [9] which states that the paper was presented before the Paduan Academy on June 21 and July 10, 1831, but it is unlikely that the volume containing the description could have reached Henry in the same year. There is strong reason to doubt whether even a brief account of the

[8] We have followed Henry's description of the machine. This is not identical with another preserved at Princeton, where the two perpendicular magnets are replaced by a single permanent magnet laid parallel to the electromagnet.

[9] *Nuovi Saggi della Imperiale Regia Accademia di Scienza, Lettere, ed Arti in Padova*, 1831. Vol. III.

machine was included in the original paper, for no other novel ideas figure in it. A long review of the paper which appeared in the *Annali delle Scienze del Regno Lombardo* in the July–August issue of 1831, makes no mention of the motor, which would have been the most striking feature of the entire paper. The probability is that Dal Negro edited his paper for inclusion in the Academy's annual publication, and we have no evidence how long after the paper was presented that an account of it was issued from the press. The first full account of this motor appeared in the *Annali delle Scienze,* March–April, 1834. Henry appears to have undisputed claim to priority in the invention.

Thomas Edmundson of Baltimore was the first to announce a modification of Henry's machine in order to perform rotary motion. He made a crude commutator to interrupt the flow of current long enough to permit each armature in turn to pass out of the magnetic field. An account of his achievement was published in 1834, the same year in which Thomas Davenport, using magnets reproduced with Chinese fidelity from one Henry had made for the Penfield Iron Works, built his own version of a rotary electromagnetic motor.

The simple little machine caught the eye and imagination of the young English physicist, James Prescott Joule, who was also making a name for himself. In his "Historical sketch of the rise and progress of Electro Magnetic Engines for Propelling Machinery" [10] he paid an early tribute to Henry's accomplishment: "The improved plan by Professor Henry of raising the magnetic action of soft iron, developed new and inexhaustible sources of force which appeared easily and extensively available as a mechanical agent; and it is to the ingenious American philosopher that we are indebted for the first form of a working model of an engine upon the principle of reciprocating polarity of soft iron by electro-dynamic agency."

Henry realized, however, that the power of the steam engine was derived from the combustion of fuel. So far from yielding to the inventor's intoxication, he carefully calculated the relative advantages of the galvanic battery and coal as agents for producing power, and concluded that his machine could have no application where economy of operation was a primary consideration.

The first technical comparison of the costs of the consumption of zinc and coal was given by Rankine.[11] He estimated one pound of

---
[10] *Annals of Electricity,* 1839. Vol. III, p. 430.
[11] *The Steam Engine and other Prime Movers.* London, 1859, p. 541.

zinc could not produce more than one-tenth the energy in a pound of coal; and while the efficient utilization of the energy was four times greater, its useful work was less than one-half that of coal, while its cost was 40 to 50 times greater. This calculation was made at a time when costs and efficiencies were at much the same stage as they were in Henry's day, and it may be assumed that these considerations influenced Henry's opinion of his "philosophical toy."

Years later, in 1876, Henry recalled this little machine in a letter to a friend of Princeton days, Rev. Samuel B. Dod: "I never regarded it as practical in the arts because of its great expense of power, except in particular cases where expense of power is of little consequence." [12]

While Henry was a professor at Princeton, a maker of what was then called "philosophical apparatus" called upon him to display his merchandise, among which was a version of this elementary machine. Henry thereupon showed his visitor his own early model of the apparatus. The dealer examined it critically, disparaged the workmanship, and admonished the professor that he should make no more models of that kind under penalty for infringing the dealer's patent rights!

The study of electromagnetism and the construction of magnets did not absorb the whole of Henry's leisure. In the month of September 1830, Professor James Renwick of Columbia College had asked him to determine the magnetic intensity of the earth at Albany. With the aid of his brother-in-law, Stephen Alexander, Henry made daily observations over a period of two months. His magnetic needles "were suspended according to the method of Hansteen in a small mahogany box, by a single fiber of raw silk. The box was furnished with a glass cover, and had a graduated arc of ivory on the bottom to mark the amplitude of the vibrations." [13]

Sometime about October the observations were suspended to enable Henry to collect his notes and to prepare the paper on his August experiments. The observations were not resumed until the following April. At noon on April 19, the oscillations of the needles were checked and found to be normal, but at six o'clock Henry noted that the needles indicated an abnormally high degree of intensity. Doubt-

[12] *Memorial,* p. 143.
[13] *Amer. Jour. of Sci.* April, 1832. Vol. XXII, p. 145.

less this increase aroused speculation in the mind of the observer, but when darkness had fallen the solution was presented to him. The inhabitants of northern towns were enthralled by an exceptionally brilliant display of the aurora borealis, which was drawn to Henry's attention at about nine o'clock.

Had he had access to a large collection of scientific literature, he might have known that the correspondence between the aurora and the disturbance of magnetic needles had been observed by John Canton a hundred years earlier. Franklin and Wargentin had also observed the association, so that it is possible Henry knew what to expect. When the glory of the heavens had reached its climax at about ten o'clock, he walked over the Academy grounds to observe the indication of the needles, and found that they registered an abnormally low degree of magnetic intensity.

At this time the Northern Lights were regarded by many in the same category as comets and earthquakes—messages from an unseen world, and most of them omens of impending disaster. Not until 1879, when Loomis finally associated the aurora with sunspot activity, was the source of the disturbance known. Before this date a wide field was open for speculation and investigation. Records by competent observers were welcomed.

Henry began in his usual fashion by checking the other men's observations, and when an English journal arrived in Albany, he found to his joy that the brilliant aurora of April had been observed in England. This suggested to him a comparison of English and American observations of earlier recorded displays of the phenomenon. He searched through the *Annals of Philosophy* for records on the other side of the Atlantic and compared the dates with records published by the meteorological stations of the State of New York. This comparison made evident the fact that the aurora borealis was no local manifestation.

On January 26, 1832, he read a paper before the Albany Institute which was published in Silliman's *Journal*.[14] He expressed the opinion: "these simultaneous appearances of the meteor in Europe and America would therefore seem to warrant the conclusion that the aurora borealis cannot be classed among the ordinary local meteorological phenomena, but that it must be referred to some cause connected with the general physical principles of the globe; and that the more

[14] *Amer. Jour. of Sci.* April, 1832. Vol. XXII, p. 143.

energetic action of this cause (whatever it may be) affects simultaneously a greater portion of the northern hemisphere."

He must have derived a peculiar satisfaction from this conclusion, for he held all through his life a firm conviction that investigation would reveal that many unsuspected relationships could be determined, and that by adding them together he would arrive at a conception of orderliness and unity in the universe. Physical phenomena of all kinds, he held, were to be regarded as forming a vast interrelated web, governed by some unknown but simple laws. Discovery of the relationships was only to be achieved by reconstructing the web fragment by fragment, atom by atom. He liked to think each new discovery, even though it were a very small one like this of the aurora borealis, provided a closer approach to the fundamental universal laws, and to their Author.

He was busy with other meteorological studies during the year 1831; in that year he compiled the tabular statement of the latitudes, longitudes, and altitudes of the forty-two observation stations in the state, to which allusion has already been made. In addition, he prepared the annual abstracts of the periodical reports of the observers. Although of small intrinsic value these details of his occupations deserve our attention. They contribute to form a picture of a busy man, and they help to convey some indication of the small amount of leisure left to him in which he could speculate upon the nature of the electrical experiments he would undertake when August came around with its opportunity to resume laboratory work.

If Henry had been inclined to regard his quantity magnet as scientifically more important than the intensity magnet, the inclination would have been caused by the demands made upon him to construct examples of the former type. These magnets were sought by other men engaged in teaching science, who wanted powerful magnets for their own use. After Silliman, one of the first to make such a request was Professor Parker Cleaveland of Bowdoin College, who, like many others, was fearful of the cost of electrical apparatus. He wanted to know how much the Yale magnet had cost. On being informed that it cost $40, Cleaveland ordered one.

In their correspondence we arrive at an understanding of the difficulties under which Henry was laboring. "I can get nothing done in the philosophical line in Albany," he laments, apologizing for delays, "except that I stand over the workmen or, which is most often the case, do the work myself. . . . The winding of the wire was done with great care and under my constant inspection. The process was a very tedious one and occupied myself and two other persons every evening for two weeks." [15]

The "two other persons" may have been Stephen Alexander and Philip Ten Eyck, for the latter had done much of the wire-winding for the Yale magnet. The construction of the Bowdoin magnet involved the winding of a thousand feet of wire.

But this operation dwindles into insignificance alongside another project he had undertaken. Again we are compelled to lament the fact that Henry did not bequeath to us a diary comparable to that of Faraday, to which he confided the reasons for undertaking his projects. All we have in this case is a hint that he was contemplating something on a very large scale, with never a word to indicate what was in his mind. That it was something sensational may be gathered from the fragments of his correspondence at the time.

In a letter dated November 6, 1831, addressed to Professor Cleaveland, Henry said: "I have lately had forged a large horseshoe weighing 101 pounds which I intended fitting up for some contemplated experiments in the identity of electricity and magnetism. It will be more powerful than any hitherto made, being almost double the weight of the Yale College magnet, (59½ lbs.)." [16]

Apparently, news of this monster magnet had spread abroad, for Silliman at New Haven learned of it and, scenting a story behind it, wrote to Henry (March 5, 1832): "I have been told that you have constructed a larger galvanic magnet than those of which notices have been published. I write to inquire if there is anything you would like to have published in the *American Journal* for April as to any extension or new application of your discovery. I am informed that you have applied it to the separation of fragments of magnetic iron from the mixture of other things." [17]

The last sentence does not refer to any of Henry's own work, but to the industrial application of the magnet by the Penfield Iron Works.

Henry put off answering Silliman's letter for three weeks, and then (March 28, 1832), he wrote:

"In answer to your inquiries as to my further experiments in

[15] Henry, Mary A. America's part in the discovery of Electro-Magnetism. *Electrical Engineer*. Jan. 20, 1892. Vol. XIII, p. 154.
[16] *Ibid.*, p. 54.
[17] *Ibid.*

magnetism, I have little definite to communicate at the present time. I have not entirely abandoned the subject, but have been prevented by circumstances from prosecuting a series of experiments, which were commenced on a very extensive plan. I had partially finished a magnet much larger than any before made, and have constructed a kind of reel on which more than a mile of copper wire was wound. I was obliged to abandon the experiments on account of the room in which my apparatus was erected being wanted for the use of the Academy and it was not convenient for me to resume them during the winter."

A kind of reel on which a mile of wire was wound! And a larger magnet than any before made! Here is a tantalizing problem. What were they for? Some students of Henry's work, including Crowther and Cajori [18] are of opinion that he had begun to construct a dynamo.

Nothing in the whole story of Henry's life and work is so tantalizing as the gaps in his activities during this year of 1831. Obviously, some quite exceptional project was in contemplation. It never was resumed. That it had to do directly or indirectly with the identity of electricity and magnetism is clear from the phrase in his letter to Cleaveland, but the purpose is obscure. Super-magnet and mile-long reel vanished into the confusion created by the arrival in Albany of copies of the *Library of Useful Knowledge* or of the *Philosophical Magazine*. When Henry read the first of these to arrive, his little world must have swayed unsteadily under his feet.

The particular paragraph that contained stunning news was in a note which Peter M. Roget attached to an article appearing in the 117th number of the *Library of Useful Knowledge,* as follows:

> *Note.* Since the above was sent to the press, a paper, by Mr. Faraday, has been communicated to the Royal Society, disclosing a most important principle in electro-magnetism, of which, I regret, I can give only the following brief statement.
>
> By a numerous series of experiments, Mr. Faraday has established the general fact, that when a piece of metal is moved in any direction, either in front of a single magnetic pole, or between the opposite poles of a horseshoe magnet, electrical currents are

---

[18] Crowther, J. G. *Famous American Men of Science,* 1937, p. 181; Cajori, F. *History of Physics,* 1938, p. 248.

developed in the metal, which pass in a direction at right angles to that of its own motion. The explication of this principle affords a complete and satisfactory explanation of the phenomenon observed by Arago, Herschel, Babbage, and others, where magnetic action appears to be developed by mere rotary motion, and which have been erroneously ascribed to simple magnetic induction and to the time supposed to be required for the progress of that induction. The electro-magnet effect of the elective [sic] current induced in a conductor by a magnet pole, in consequence of their relative motion, is such as tends continually to diminish their relative motion; that is, to bring the moving bodies into a state of relative rest: so that, if the one be made to revolve by an extraneous force, the other will tend to revolve with it, in the same direction, and with the same velocity.

Roget's article is dated December 12, 1831, so that it is improbable Henry would see it before the following March. The dates are significant in view of the claim that Henry had discovered the principle of induction before Faraday.

The next mention of Faraday's success in attaining mutual induction appeared in the *Philosophical Magazine* for April 1832, in these words:

> Feb. 17 — Mr. Faraday gave an account of the first two parts of his recent researches in electricity; namely volta-electric induction, and magneto-electric induction.
>
> If two wires A and B, be placed side by side, but not in contact, and a voltaic current be passed through A, there is instantly a current produced, by induction, in B, in the opposite direction. . . . If a wire, connected at both extremities with a galvanometer be coiled, in the form of a helix, around a magnet, *no current* of electricity takes place in it. . . . But if the magnet be withdrawn from or introduced into such a helix, a current of electricity is produced *whilst the magnet is in motion,* and is rendered evident by the deflection of the galvanometer. . . . Thus is obtained the result so long sought after—the conversion of magnetism into electricity.[19]

[19] *Phil. Mag.* April, 1832. Vol. XI, p. 300.

THE DISCOVERY OF INDUCTION

It would be difficult to analyze Henry's feeling when he read the words intimating that Faraday had succeeded in a quest wherein Henry thought he might gain the prize.

Ever since Oersted had rediscovered what Romagnosi had pointed out, that a wire bearing an electric current would deflect a magnetic needle, many experimenters had groped for the discovery of the reverse effect, the production of a current in a wire through the medium of magnetism. Both Faraday and Henry were persuaded that patient investigation would reward them with proof that magnetoelectricity could be created with the aid of the electromagnet. The English scientist had begun experiments toward this end in 1824, while Henry was still a student, but he laid aside his investigations in the belief that his magnets were too weak to furnish appreciable results.

The lives of Faraday and Henry ran in strange parallels. The former was born in 1791 and died in 1863; Henry in 1797 and 1878, so that their productive years were practically simultaneous. The two men sprang from the same working class that had no traditions of knowledge and that did not possess the means of educating its sons. Both were influenced to follow a scientific career by the chance reading of books which revealed the works of nature to their imaginations. They were both deeply religious men, who regarded Nature as the handiwork of their Creator.

But no comparison of the two would be complete without reference to their relative situations at this stage of their careers. We have seen Henry at work in a small town, with only very little time, hampered for want of funds and facilities, with few technical companions. Faraday was much more favorably situated at the Royal Institution, living in the world center of intellectual activity, with as much time as he wanted for research, in constant communication with some of the most brilliant technical minds of the day and, if the Royal Institution was not so wealthy as it had been in Davy's time, there was the well-furnished laboratory he had built up when funds were plentiful.

Primarily interested in chemistry, Faraday had been experimenting since 1824 with the voltaic battery, convinced that if electricity could produce magnetism, then a magnet must somehow produce electricity. His early experiments all had negative results. Strange to say, one of these experiments should have produced encouraging results.

THE DISCOVERY OF INDUCTION

Passing a galvanic current through a wire lying close to another connected with a galvanometer, he observed no result, when of course he should have witnessed a reaction. Either his battery was too weak or his galvanometer was not sensitive.

He resumed his experiments in 1828 with no better fortune. Then, in 1831, he learned about the magnets of Moll and Henry, and began all over again with stronger magnets than any he had previously possessed. It is singular that he does not mention the names of either Moll or Henry in these early papers, but there is no doubt that he had read of Henry's work and had profited by it, as will be seen from the description of the magnet he had constructed.

Faraday had made a ring of soft iron which he wound with many coils of wire. Why did he chose a ring? Whether by accident, design, or impulse, he does not say. The inference is that he was influenced by Henry's description of his large magnet with an armature joining the poles.

He made two separate coils on this ring magnet knowing that if he passed a current through one coil it would magnetize the iron core, and he was curious to learn what happened to the second coil. The latter was connected to a galvanometer. Once he closed the battery circuit he observed that the galvanometer needle was deflected. "It oscillated and settled at last in the original position." [20] No continuous effect was noted. He broke the battery circuit and observed an identical effect, except that the needle swung in the opposite direction.

He was somewhat perplexed, but was by no means convinced he had discovered evidence of what he was seeking. He tried other tests, using first those chemical tests with which he was most familiar. These failed him.

He tried other arrangements, some of which gave more promising results. He experimented with alternatives to opening and closing the circuits in order to supplement the one shred of evidence that pointed in the right direction, but he derived no encouragement from them. He even brought a permanent magnet up to the ring and, surprising to relate, he observed no reaction. More experiments were performed with the ring magnet on August 29, but with no better result.

[20] Jones, B. *Life and Letters of Faraday*. Vol. II, p. 3; quoted from Faraday's notebook.

Two weeks later, on September 12, he rearranged his apparatus and began all over again. Now he got a glimpse of the truth, but was reluctant to decide that it was conclusive. On the eve of the 24th, he wrote to a friend, "I am busy just now on electro-magnetism, and I think I have got hold of a good thing but can't say. It may be a weed instead of a fish, that after all my labor, I may at last pull up." [21]

In deflecting the galvanometer needle the electric current has to overcome both the force of inertia and the force of terrestrial magnetism, which tries to return the needle to its normal position of rest. In order to free the galvanometer needle from the action of the earth's magnetism it was necessary to use an astatic needle. Nobili had given a full description of the astatic galvanometer in 1825; although the distinction between the throw and steady deflection was not understood, the superior sensitivity of the instrument was clear.

Faraday does not appear to have recognized the need for an astatic needle, for he more than once failed to observe effects that should have been apparent. Henry had a slight advantage of him in respect to apparatus, for, as we shall see, he grasped the effects that eluded Faraday. When the latter did observe the very thing he was looking for, when he saw the galvanometer needle oscillate on making and breaking the circuit, he failed to grasp its significance. Cautiously, he asked himself: "May not these transient effects be connected with causes of difference between the power of metals at rest and in motion, as in Arago's experiment?"

On the eventful September 24th, when he was to learn whether he had hooked a fish or a weed, Faraday once more rearranged his apparatus, dispensing with the ring magnet and substituting a bar magnet. With this he performed the conclusive experiment of passing the bar in and out of a helix. At last he had the satisfaction of observing that electricity was created when the bar was in motion. He made some additional experiments in confirmation before assembling the data which he presented to the world on November 24, 1831, two months after the crucial experiment had been performed.

Let us once more draw attention to the fact that in these ten days of experiment, Faraday more than once had the necessary evidence at his finger-tips but the solution eluded him. The truth of the matter seems to be, that his succession of experiments were correctly conceived to establish proof of his hypothesis. He was foiled in his search by the

[21] *Ibid*.

insensitivity of his instruments. His thinking was accurate, but he distrusted his mind more than he distrusted his instruments.

The preliminary report of Faraday's achievement stimulated Henry to write an account of what he had been doing in the same direction. They were but two of the numerous investigators working on this problem.

The failure of the acutest minds to find the long-suspected connection between electricity and magnetism until twenty years after the production of a steady electric current is some indication of the difficulty attending this work. Henry had been occupied with the problem, but how far he had progressed it is now impossible to say. His manifold occupations had not left him much time for the preparation of scientific papers. In this respect he was greatly behind his British rival who, by 1831, had already published sixty papers on his researches. But the publication of Faraday's announcement of his discovery was too important for Henry to overlook without telling the world of his own performance.

He dashed off an article for Silliman's *Journal*. The July number was already in print when the editor received it, but Silliman instantly perceived its significance to American science, and tacked it onto the July issue as a sort of appendix, under the title of "On the Production of Currents and Sparks of Electricity and Magnetism." [22]

In speaking of the paucity of information on such an important scientific event as Faraday's discovery, Henry grumbles with unconscious humor, "No detail is given of these experiments, and it is somewhat surprising that results so interesting, and which certainly form a new era in electricity and magnetism should not have been more fully described before this time in some of the English publications."

This is an odd attitude to be adopted by a man who is about to announce for the first time experiments he had performed before the victim of his reproach had performed them.

Henry's version of the affair is best shown in his own words. Having commented upon the failure of other attempts to produce the desired result, he says:

> It early occurred to me that if galvanic magnets on my plan were substituted for ordinary magnets, in researches of this kind, more success might be expected. Besides their greater power,

[22] *Amer. Jour. of Sci.* July, 1832. Vol. XXII, p. 403; *S.W.* Vol. I, p. 73.

these magnets possess other properties, which render them important instruments in the hands of the experimenter; their polarity can be instantly reversed and their magnetism suddenly destroyed or called into full action, according as the occasion may require. With this view, I commenced last August the construction of much larger galvanic magnet than, to my knowledge, had been before attempted, and also many preparations for a series of experiments on a large scale, in reference to the production of electricity from magnetism.

I was however at that time accidentally interrupted in the prosecution of these experiments, and have not been able to resume them, until within the last few weeks, and then on a much smaller scale than was at first intended.

He then recounts how he had been anticipated by Faraday, and how he had learned of the British scientist's methods. It is well to remember that the only experiments he had learned of Faraday performing were those in which a magnet had been moved in a helix, and that when a piece of metal was moved in front of a magnetic pole electric currents were developed.

He then proceeds:

Before having any knowledge of the method given in the above account, I had succeeded in producing electrical effects in the following manner, which differs from that employed by Mr. Faraday, and which appears to me to develop some new and interesting facts.

A piece of copper wire about thirty feet long and covered with elastic varnish [23] was closely coiled around the middle of the soft iron armature of the galvanic magnet, described in volume xix of the *American Journal of Science,* and which when excited will readily sustain between 600 and 700 pounds. The wire was wound upon itself so as to occupy only about one inch of the length of the armature, which is seven inches in all. The armature thus furnished with the wire, was placed in its proper position across the ends of the galvanic magnet and then fastened so that no motion

[23] This form of insulation had been described by Hare in Silliman's *Journal* of July 1831. "I use no wrapping, but merely shell lac varnish applied to the winding." Prof. J. W. Webster of Harvard says in the same issue: "In constructing the magnetic apparatus, there is considerable economy in using sealing wax instead of silk."

THE DISCOVERY OF INDUCTION

could take place. The two projecting ends of the helix were dropped into two cups of mercury, and connected with a distant galvanometer by means of two copper wires, each about forty feet long.

This arrangement being completed, I stationed myself near the galvanometer and directed an assistant at a given word to immerse suddenly, in a vessel of dilute acid, the galvanic battery attached to the magnet. At the instant of immersion, the north end of the needle was deflected 30 degrees to the west, indicating a current of electricity from the helix surrounding the armature. The effect however appeared only as a single impulse, for the needle, after a few oscillations, resumed its former undisturbed position in the magnetic meridian, although the galvanic action of the battery, and consequently the magnetic power was still continued. I was however much surprised to see the needle suddenly deflected from a state of rest to about 20 degrees to the east, or in the contrary direction, when the battery was withdrawn from the acid, and again deflected to the west when it was re-immersed.

This operation was repeated many times in succession, and uniformly with the same result, the armature the whole time remaining immovably attached to the poles of the magnet, no motion being required to produce the effect, as it appeared to take place in consequence of the instantaneous development of the magnetic action in one, and the sudden cessation of it in the other.

The experiment illustrates most strikingly the reciprocal action of the principles of electricity and magnetism, if indeed it does not establish their absolute identity. In the first place, magnetism is developed in the soft iron of the galvanic magnet by the action of the currents of electricity from the battery, and secondly, rendered magnetic by contact with the poles of the magnet, induces in its turn currents of electricity in the helix which surrounds it; we have thus as it were electricity converted into magnetism and this magnetism again into electricity. . . .

From the foregoing facts, it appears that a current of electricity is produced, for an instant, in a helix of copper wire surrounding a piece of soft iron whenever magnetism is induced in the iron; and a current in an opposite direction when the magnetic action ceases; also that an instantaneous current in one or the other direction accompanies every change in the magnetic intensity of the iron.

Although we are here concerned with the discovery of mutual induction, the reader should observe that, at this point, Henry had actually built the first iron core transformer, although he did not think of it as such.

It is clear from what Henry quotes of his knowledge of Faraday's work (and as we know, he had a very incomplete account of it) that he felt the emphasis was laid upon the motions of the magnet when the induced currents were produced, and that he thought his experiments, in which the magnet did not move, were distinctly different from the others.

Faraday had performed experiments with his ring magnet similar to those Henry described in the making and breaking of the circuit. The essential apparatus, the means of discovery, and the observed phenomena were almost identical. Yet how different was the individual perception.

Why should Henry instantly have recognized this transient effect for what it signified, and Faraday distrust it? The latter suspected it might be due to the difference between the powers of metals at rest and in motion.

Probably because Henry had given more thought directly to Oersted's experiment and had more accurately deduced its nature, while Faraday took a more general view of his problem. Electric current is moving electricity, a dynamical phenomenon, while the deflection of the magnet is a static phenomenon. Stationary magnetism cannot produce a steady current of electricity because it does not belong to the same type of phenomena. If moving electricity produces magnetism, then motion associated with magnetism must be required to produce electric current.

How different was the flash of recognition with which Henry perceived the kick of the needle. Like Faraday, he had reasoned how the desired effect might be obtained. He had developed the tools which would enable him to produce a sufficient amount of electrical energy for conversion and reconversion. He anticipated a loss of energy in the process and, so, had provided an ample store. Furthermore, his vision was unclouded by previous failures.

THE DISCOVERY OF INDUCTION

This discovery of electromagnetic induction marked the beginning of the modern era of electricity, one is tempted to write, of modern physics. It underlies every dynamo, alternator, and electric motor that has been put into use.

Henry never publicly claimed to have discovered mutual induction, but in private he expressed regret that he had delayed announcing the results of his experiments which would have enabled him to establish his right to the discovery. That there was delay in publishing an account of the experiments made after August 1830 cannot be doubted. Lack of time may be the excuse, since we have tried to describe his many occupations at the time. But we cannot rid ourselves of the suspicion that he possessed the proofs had he cared to use them.

The reader will agree, I think, that no one in Henry's straitened circumstances would have gone to the cost and labor of making the largest magnet in existence to perform an initial experiment. He would be much more likely to have performed a tentative test with what magnets he had before embarking upon the construction of more elaborate apparatus. What he seems to have had in view was proof upon the most impressive scale, which is confirmed in that passage of his article where he writes of ". . . preparations for a series of experiments with it on a large scale." This would be quite in keeping with his character.

In the first place, his caution against hasty conclusions required that he make no decision until he could provide substantial proofs. The object of his search was of unusual importance, and he may have felt the need for more richly illustrating the bare discovery of a force so ardently sought. He was still young. The romance of the theatre still clung to him. He had repeatedly yielded to the employment of theatrical methods to obtain demonstrative results. He was something of a showman in this respect.

Having received as yet small recognition in Europe as the creator of the powerful electromagnet and being on the verge of announcing a notable discovery, he seems to have asked himself whether he should permit history to judge him as one who fidgeted with gadgets, or should he burst upon the world with *éclat?*

The planned dramatic flourish was a fiasco.

The question whether Faraday or Henry was the first to discover electromagnetic induction does not appear to admit an absolute answer. Henry's claim is not devoid of substance, and there can be no denial he made a definitive experiment entirely on his own.

The strongest advocate of Henry's claim to priority was his daughter Mary, who set forth his claim in a series of articles.[24] Making every allowance for the fact that Mary Henry was the recipient of her father's confidences, we feel reluctant to accept without superior documentation, a conclusion that does not fit neatly into the situation.

Drawing upon her personal recollection, Miss Henry wrote:

> Often by the fireside in his home, he told to his wife and daughters the story of his great early disappointment. How he made the discovery of the extra current five years before Faraday. How the discovery of magnetoelectricity followed in 1830. How eager he was to pursue it, and how baffled for lack of time and materials. How he wished to amplify his results before he gave them to the world, and how he commenced his great preparations for this purpose, and then one day in the library of the Academy, seizing eagerly upon a newly-arrived periodical, how he came suddenly upon the notice which told him, that although he had made the discovery so long before anyone else, Faraday must claim it.

Miss Henry claimed that her father had made the discovery of magnetoelectricity before Faraday, asserting that he had detected a spark produced by it as early as 1829. She quotes a letter written by Henry's assistant, George W. Carpenter. Unfortunately no date is attached to this letter, and the suspicion attaches to it that it was written long after the incident described. However, he wrote:

> In a well-remembered conversation with me he [Henry] alluded to an incident in his own experience. After retiring one night he worked out mentally how he could probably draw a spark from the magnet. Upon rising in the morning he hurried to his working room, arranged the apparatus, tried the experiment, when success crowned his labor. He had accomplished what had never been done before. Unfortunately, he failed to

[24] *Electrical Engineer.* Vol. XIII, 1892.

publish his discovery. In continuing his remarks, he added that Faraday, the great European philosopher, some time after, successfully tried the same experiment and at once announced it, unfortunately before Professor Henry's success was publicly known. Mr. Henry regretted that through his own negligence he had been deprived of the honor awarded to Faraday.

Memory plays strange tricks upon us at times. It seems incredible that Carpenter who was working with Henry during 1830 and 1831 should have known nothing about the spark until Henry informed him in a casual conversation. Henry, and presumably Carpenter, was familiar with electric sparks, he was too well informed upon their significance to have allowed this unusual appearance to pass without comment in his earliest paper on the magnets. Cautious and careful he may have been, but not to this extent.

In point of fact, Henry discounts this evidence by a statement in this very paper in which he announces his independent discovery of electromagnetic induction. He wrote: "I have been anticipated in this experiment of drawing sparks from the magnet by Mr. James D. Forbes of Edinburgh, who obtained a spark on the 30th of March; my experiments being made during the last two weeks of June.... My result is entirely independent of his and was undoubtedly obtained by a different process."

While this eliminates any doubt about the observation of the spark in 1829, it does not dispose of other evidence.

Miss Henry quotes another witness, Rev. Theodore L. Cuyler, who wrote her as follows: "Your father often spoke to me of his disappointment about that discovery. 'I ought to have published earlier,' he used to say. 'I ought to have published, but I had so little time and how could I know that another man on the other side of the Atlantic was busy with the same thing?'" [25]

The actual date of Henry's critical experiments cannot be fixed, but there is still justification for supposing that he had anticipated his rival, although unaccountably laggard in announcing his success.

Our inclination is to believe that he made the all-important observation of the galvanometer needle sometime in August 1830, too late for him to amplify his experiment or to check it by other methods. Because of the absence of corroborating evidence he omitted mention

[25] *Ibid.*, Vol. XIII, p. 153.

of it in his description of his magnets in Silliman's *Journal* in the following January. The belief, unsupported by any direct evidence, is supplemented by the further belief that the ensuing year was devoted to much thought and no little experimenting on a minor scale. When August arrived he was sure of himself and the discovery he had made. All he required was the means of making a spectacular demonstration that would be conclusive.

This would appear to be confirmed by his own words: "I commenced last August the construction of a much larger galvanic magnet than, to my knowledge, had ever before been attempted, and also made preparations for a series of experiments on a larger scale, in reference to the production of electricity from magnetism. I was however at that time accidentally interrupted in the prosecution of these experiments and have not since been able to resume them, until within the last few weeks, and then on a much smaller scale than was at first intended." [26]

The accidental interruption was caused by the resumption of school duties after the holidays. The one precious month in the year in which original experiments could be made had been devoted to the laborious task of winding coils on the mammoth magnet and that reel on which more than a mile of copper wire had to be wound. It is inconceivable that he would sacrifice that whole month to this work unless he had an unusually important object in view. The object was stated in his letter to Cleaveland, dated November 16, 1831, before he had any knowledge of Faraday's success. It was to establish "the identity of electricity and magnetism."

One other piece of circumstantial evidence pointing to priority in the discovery resides in the nature of the experiment by which Henry arrived at his conclusion. He was wholly ignorant of Faraday's iron ring experiment. He was justified in supposing that his method differed from that of Faraday. The procedure both had followed in performing this experiment was the most logical in the state of knowledge at the time. Henry had a better practical understanding of the principles underlying the relationship between core, battery, and coils. His recognition of the result of the experiment, which Faraday had distrusted and rejected, was natural.

We have not labored this point of priority to disparage Faraday's achievement or to award the honor to Henry. There is enough honor

[26] *Amer. Jour. of Sci.* Vol. XXIII, p. 403; *S.W.* Vol. I, p. 73.

in the discovery for both men. But if Henry's share in the discovery is not mentioned in the histories of the science, where it is accredited to Faraday, such an important biographical incident cannot be glossed over in a study of his work.

That Henry felt the pangs of disappointment at having the honor snatched from his grasp is shown in the recollections of his daughter. None the less, his generous nature never permitted him to contest the fact that Faraday, by prior publication, had established claim to precedence. So shy a man would hardly have given away his woes, even to a diary.

Personal disappointment may not constitute the main factor in Henry's disappointment. In the first half of the nineteenth century American science counted for little in the world. Had Henry been able to secure priority in the discovery of mutual induction, the gain to American science would have been immense. The originality of the discovery constituted such a signal achievement, outweighing his later accomplishments in the electrical science, that it could have given American scientists an inspiration and would have gained a more worthy respect for their efforts.

The question of priority is of little importance now; the significant fact is that the phenomenon of electromagnetic induction was discovered independently and at about the same time in London and in Albany.

Though Henry lost his chance to establish himself as the discoverer of mutual induction, the last paragraph of the long-delayed paper of 1832, almost an afterthought, reveals Henry's greatest single contribution to science, the discovery of electromagnetic self-induction, or "extra current" as it was called at first. The passage goes:

> I have made several other experiments in relation to the same subject, but which more important duties will not permit me to verify in time for this paper. I may however mention one fact which I have not seen noticed in any work, and which appears to me to belong to the same class of phenomena as those before described; it is this: when a small battery is moderately excited by diluted acid, and its poles, which should be terminated by cups of mercury, are connected by a copper wire not more than a foot in length, no spark is perceived when the connection is either

formed or broken; but if a wire 30 or 40 feet long be used instead of the short wire, though no spark will be perceptible when the connection is made, yet when it is broken by drawing one end of the wire from its cup of mercury, a vivid spark is produced.

If the action of the battery be very intense, a spark will be given by the short wire; in this case it is only necessary to wait a few minutes until the action partially subsides, and until no more sparks are given from the short wire; if the long wire be now substituted a spark will again be obtained. The effect appears somewhat increased by coiling the wire into a helix; it seems also to depend in some measure on the length and thickness of the wire. I can account for this phenomena only by supposing the long wire to become charged with electricity, which by its reaction on itself projects a spark when the connection is broken.

If any additional proof were needed that Henry was an exceptionally keen observer it lies in this passage. For thirty years since Volta's cell was built, investigators had been experimenting with currents in wires of different lengths. No one had commented upon the observation of sparks occurring when a long circuit was broken. It is improbable that no one had seen a spark. There must have been some such spark in Barlow's experiments with the projected telegraph, in some of Sturgeon's experiments, and possibly in Ampère's experiments with coils of different shapes and sizes. We have already mentioned Pouillet's disconcerting experience.[27]

Henry was first to comment upon the presence of these sparks and to place an accurate interpretation upon their cause. The observation of the spark is not of itself remarkable, but the immediate recognition of its nature is a truly remarkable achievement. The phenomenon of self-induction was not even suspected, which makes the achievement all the more noteworthy.

In this phenomenon electric current behaves, on account of self-induction, like a current of water in a pipe. When it is set in motion a slow build-up occurs—in the case of water because of mechanical inertia, and in the case of electric current as a consequence of electromotive force. When the current is interrupted we perceive a tendency to flow on—again on account of self-induction in the wire, and inertia in the case of water.

[27] See footnote 6 of this chapter.

The phenomenon of inertia in the instance of the current must not be interpreted as inertia of electricity in the wire, for the magnitude of the self-induction is very different in the same wire according to the method of its winding and to its surroundings. It is much larger in a coil of wire, as Henry observed, than in the same wire stretched out straight, and it becomes much greater if an iron core is introduced, as he observed in later experiments.

When did Henry discover self-induction? His daughter and Dr. Ames [28] say it occurred in 1829. Henry lends support to this date (or perhaps it was 1830) in his report to the Regents of the Smithsonian Institution when he felt called upon to defend his scientific reputation against Morse's assault. He wrote: "Indeed my experiments on the transmission of power to a distance were superseded by the investigation of the remarkable phenomenon which I had discovered in the course of these experiments, of the induction of a long wire upon itself, and of which I first made mention in a paper in Silliman's *Journal* in 1832." [29]

This paper of Henry's seems to have escaped Faraday's attention, for in 1834 he announced that he had made the identical observation of self-induction as though it were an original discovery.

Few scientific papers of its length have contained the announcement of two such important discoveries. Upon these discoveries were laid the foundations of the century's far-reaching inventions embracing electromagnetism. The recognition of Henry's claim as an independent and parallel discoverer of mutual-induction is frequently omitted from histories of science, but his discovery of self-induction is freely conceded. Not until 1886, after Henry's death, when the Smithsonian Institution published Henry's scientific works was Henry's labor revealed to many Englishmen who probably had never seen a copy of Silliman's *Journal*. At that time some handsome tributes were paid him to atone for a half-century of neglect. For example, J. A. Fleming, professor of electrical engineering at University College in London, wrote:

> At the head of this long line of illustrious investigators stand the preëminent names of Faraday and Henry. On the foundation stones of truth laid down by them all subsequent builders have

[28] *Discovery of Induced Currents.* Vol. I, p. 9.
[29] Smith. Inst. *Annual Report for 1857*, p. 106.

been content to rest. The *Electrical Researches* of the one have been the guide of the experimentalist no less than the instructor of the student, since their orderly and detailed statement, alike of triumphant discovery and of suggestive failure, make them independent of any commentator. The *Scientific Writings* of Henry deserve hardly less careful study, for in them we have not only the lucid explanations of the discoverer, but the suggestions and ideas of a most profound and inventive mind, and which indicate that Henry had early touched levels of discovery only just recently becoming fully worked.[30]

At the International Congress of Electricians held in Chicago in 1893, composed of twenty-six representative scientists drawn from nine great countries, Henry's name was given by unanimous consent to the standard unit of inductance.[31] It is somewhat surprising that the suggestion to call this unit the "henry" had its impulse from the foreign representatives, and that no American worker in the field had thought of so honoring a fellow countryman. The proposal was made by Professor Mascart of France, and was seconded by the British Professor Ayrton. The American delegates attending Congress must have experienced a feeling that they were wanting in a sense of appreciation for their countryman's achievements. Thus Henry's name was added to that small roster of pioneers who contributed fundamentally to electrical science—Ampère, Coulomb, Faraday, Gauss, Gilbert, Ohm, and Volta. He is numbered among those workers whose names have been immortalized as electric and magnetic units.

There are several contradictions in Henry's life. Here was a man who had applied himself intelligently, diligently, and resolutely, in the face of difficulties that would have halted all but the most dauntless seekers after truth, to the solution of a series of scientific problems. He had apprehended one manifestation of truth after another. He had passed the threshold of a discovery and was exploring its interior secrets so that his name might be written in bold letters upon the pages of history—and he had been forestalled in the announcement.

---

[30] *The Alternate-Current Transformer*. London, 1889. Vol. I, p. vii.
[31] A coil has an inductance of one henry if a rate of change of current of one ampere per second through it produces an electromotive force of one volt across its terminals.

THE DISCOVERY OF INDUCTION

Henry was never the man to advance his own claims, he studiously avoided any desire for self-revelation. In the face of the great disappointment which descended upon him, another man might have been overwhelmed by a sense of frustration, and indeed Henry had lost the savor for the projected experiments with his giant magnet and the great reel of wire which had filled his thoughts during August. What these experiments were intended to be we do not know. He never spoke of their nature.

There was a period, however, when he was the prey of discouragement. The story of his distress was told by James Welling, President of Columbia College.[32] Apparently it was dispelled by the prophetic remark of William Dunlap. We have no idea how Henry came to know Dunlap, the one-eyed portrait painter, diarist, novelist, playwright, and theater manager. They may have met during those romantic days when Henry haunted the theater at Albany, and it is conceivable that Dunlap had learned something about Henry's scientific work.

According to Welling's story, Henry was depressed by the feeling that he was being deserted by his friends. Who the friends were or why they should desert him we are left to surmise. Some Albany residents may have assisted him in procuring materials for his experiments and, when the young professor withdrew all claim for his greatest discovery, may have been unwilling to support him further in making what appeared to be fruitless researches. There may have been tactless suggestions that he would be better employed if he applied his talents to something more useful and profitable, such as steamboats or locomotives. Or this proud and sensitive young man who had accepted the aid of friends may have felt that he had failed them by neglecting to publish the preliminary accounts of his discovery for so long that a rival worker had been able to anticipate him.

In any event, whatever the cause may have been, Welling told how the fret of mind was reflected on his face. Observing his melancholy, Dunlap inquired why he should be unhappy when he had every reason to be satisfied that his work was being appreciated. Henry confided to him that he was dispirited by the coolness of friends who had formerly been distinguished by the warmth of their sympathy. Dunlap listened to his recital and, when Henry had fin-

[32] Life and character of Joseph Henry. *Memorial*, p. 184.

ished speaking, he laid his hand affectionately on the younger man's shoulder and said, "Albany will one day be proud of its son!"

There were few men of merit in the country whom Dunlap did not know. His opinion counted for something among them and Henry may or may not have accepted his praise at its proper value, but he did not forget it even if Dunlap did not think the remark worthy of inclusion in his diary. This is all he had to say about the incident:

> Monday Octr. 8th, 1832. Embark at 7 O'Clock in ye Champlain Steam Bt for Albany. Meet Mr. Joseph Henry asst at Dr. Beck's Acady. He has been to Princeton, N.J. where he is elected professor. Mr. Henry is one of those men who do honor to our Nature. Men who do by the force of what is called genius, by the ardent love of knowledge and desire to obtain rise superior to circumstances which would seem to have doomed them to ignorance & vice and become luminaries as well as examples to others. . . . His name will be enroll'd with those of Franklin, Silliman, Rittenhouse & other Americans who have transmitted light from the West to the East, and from the region to which light has been travelling for ages to that whence it emanated.[33]

Henry showed that he had not lost all interest in the theater for Dunlap records that he subscribed for a copy of his *History of the American Theatre* which was then in the press.

Many of the citizens of Albany could never be expected to appreciate Henry's accomplishments. He must have realized that there were many people who knew next to nothing about him other than that he taught their sons at the Academy. He might possibly have learned of the comedy of Davenport's fruitless quest in Albany for knowledge of a magnet.

Had Henry been a native of Boston, one feels that the city would have taken him to its collective bosom; his name would have been spoken on its stirring and smelling quays. But not in Albany.

When Tom Davenport came panting down from his Vermont hills in search of enlightenment, he was disappointed to learn that Henry was absent for a few days. Unwilling to waste time on his

[33] Dunlap, William. *Diary*. New York Historical Assoc. Vol. III, p. 620.

THE DISCOVERY OF INDUCTION

errand, and confident that everyone must know where Henry had his magnets made, he opened inquiries.

> I inquired of my landlord where I would be likely to "find a powerful magnetic battery." [34] He replied that they "made such things at the tinshop for blasting rocks." And he directed me there. The tinman showed me a jeweller's shop where he was sure they "made something of the kind in the manufacture of watches." The jeweller remarked that he had no occasion for such heavy instruments about watches, but would advise me to go to the Eagle Furnace where they could "cast from any pattern." At this latter place I declined calling, and returned home.[35]

The pity was that Dunlap's words were not overheard by these people and their kindred, so that the citizens in general might have appreciated Henry's work and worth while he still walked among them. His attainment was known by too few. Of these some may have been convinced he possessed genius, others that he had an animated talent, while not a few may have wondered whether he was not hollow.

The time was past for Albany to remedy its neglect. The first of Henry's three periods of effort had ended. He had worked for six years at the Albany Academy, and the years had been fruitful in discovery.

---

[34] It is interesting to note that the apparatus was commonly known by this name, only the professedly educated people spoke of it as a magnet.
[35] Davenport, p. 51.

PRINCETON

In his thirty-sixth year, Henry was invited to occupy the chair of natural philosophy at the College of New Jersey, recently vacated by Professor Henry Vethake. He must have gone to Princeton with a feeling of elation. His appointment showed that his work was meeting with recognition, and he was entering upon a field of duty which offered him greater opportunity and more facilities to pursue research.

Still, he retained nostalgic memories of those formative years in Albany. In 1863, when the Albany Academy celebrated the fiftieth year after its founding Henry was unable to attend the ceremonies but he wrote:

> I need not say to you that it would be a source of much enjoyment to me, though not unmixed with sadness, to be present at the celebration of so important an epoch in the history of an institution with which the earlier portions of my own life were so intimately connected; to turn back, as it were, the pages of the mysterious book of latent memory, and to have presented to me the events, the objects, and the associates of years long gone by. The past and the present would, however, be mingled together in a varied picture of light and shade—of pleasure in greeting the surviving friends of former times, and melancholy in holding converse in imagination with those who have departed—of gratification in beholding the improvements which, in time, have been wrought in the city of my birth; and of sadness in view of the

changes, even for the better, which have rendered me a stranger in the home of my childhood.[1]

The responsibility placed upon a university professor is a heavy one. When Henry went to the College of New Jersey in 1832 the burden placed upon him was doubly heavy. He and every other member of the faculty were confronted with one of the most difficult situations in the history of the college.

Princeton had had a turbulent existence. The Revolutionary War had swirled about its walls, the students were driven out, and Nassau Hall was converted into a barracks for British soldiers. When the military departed they left behind something of that ferment and agitation which accompanies all invading armies. The college buildings had suffered indignity, but of much greater concern was the temper of the student body. The spirit of independence inflamed them to such an extent that they resorted to violent means of protesting against what they thought was the severity of the college discipline. Unseemly episodes disturbed the semi-ecclesiastical serenity of the campus, as when students barricaded themselves in Nassau Hall; when they felled the vice-president with a wine decanter; and when they dropped bottles of ink on the hands of professors who walked incautiously beneath their windows. In 1807 three-fourths of the students were under suspension for unruly conduct.

But by this time the situation was no longer unique. Shrivelled incomes and the indecorous behavior of students appear to have been the rule rather than the exception in colleges in the first decade of the century.[2]

The reputation of the College had dwindled because of the riotous misbehavior of the students until only about seventy remained in residence. In 1829 the College was faced with the prospect of bankruptcy. At this point matters were taken in hand by the alumni. The strict rules of conduct were relaxed, more funds were raised, and a better qualified faculty was recruited. The spirit of reform was in full flood under the admirable Dr. John Maclean when Henry was called to the chair of natural philosophy.

Dr. Maclean was enthusiastic about securing Henry for the vacant

[1] Letter to Dr. Orlando Meads, June 23, 1863. *Proc. of the Albany Acad. Semi-Centennial*, p. 66.
[2] Cheney, E. P. *Hist. of the University of Pennsylvania, 1740–1940*. Philadelphia, 1940.

chair when he was nominated by Dr. Jacob Green of Philadelphia and Dr. John Torrey of New York. But when Henry's name was laid before the board of trustees one of them asked the disconcerting question: "Who is Henry?"

Fortunately, ample evidence was forthcoming to convince the trustees that Henry was a desirable addition to the faculty. Silliman, as might be expected, was forthright in his praise: "Henry has no superior among the scientific men of the country—at least among the younger men." Renwick of Columbia was unrestrained in his praise. "He has no equal," was his unqualified judgement. Accordingly he was appointed, and went to Princeton in November 1832.

In spite of the ignorance betrayed by one of their members, the trustees of the college were generally aware of the distinction the new professor was bringing to Princeton. His salary on appointment was $1,000, but in recognizing his services as well as they were able, the trustees raised this sum to $1,200, and then to $1,500 with the use of a house which was built for him on what is now the site of Reunion Hall. The house has been moved three times but still stands on the campus.

The increases of salary may have been more apparent than real, for Professor Brackett used to say Henry told him he sometimes received no more than $600 a year.[3] Still, the position had many compensations, for he enjoyed the admiration and friendship of the faculty and the respect of the students. After the first year the pressure of his duties did not interfere with his investigations.

Henry used to declare that his fourteen years at Princeton were the happiest years of his life. Certainly they were fruitful years. Many of his accomplishments there contained the promise of greater things to come, and he was to display an interest in a broader field of scientific thought.

Henry's magnets had brought fame to his modest laboratory in the basement of Albany Academy, and now he carried some of the luster as he stood before his appreciative students. Yet, in going over the glowing accounts of his young acolytes of these years, one is more impressed by their ardor for the man than by their admiration for his work. Few college professors have left such a deep impress of his character upon their students. His substantial work at the College has caused the members of the faculty to regard him always

[3] Magie, W. F. Joseph Henry. *Rev. Mod. Phys.* Vol. III, Oct. 1931, p. 465.

as Princeton's special ornament. Such tributes from faculty and alumni are rarely accorded. The man whose character and abilities have a special appeal to students is not always held in equal respect by the faculty.

Princeton had learned about Joseph Henry without knowing him. The reputation for wisdom he had acquired no doubt caused wholly erroneous pictures of the natural philosopher to be drawn in the minds of those who awaited his coming. Great was the astonishment when the bewhiskered and grizzled object of their imaginings turned out to be a youngish man possessing a pleasing appearance and an attractive personality.

At the time of the unveiling of the Saint Gaudens memorial tablet at the university, one of his former students, Edward Dickerson recalled the surprise occasioned on Henry's arrival. That he was a man of unusually striking appearance may be gathered when the exuberant rhetoric has been pruned from the language employed in the description.

> His clear and delicate complexion, flushed with perfect health, bloomed with hues that maidenhood might envy. Upon his splendid front, neither time nor corroding care, nor blear-eyed envy, had written a wrinkle nor left a cloud; it was fair and pure as monumental alabaster. His erect and noble form, firmly and gracefully poised, would have afforded to an artist an ideal model for Apollo. The joy of conflict and triumph beamed from his countenance—a conflict in which for years he had struggled with the phantoms that guard the hidden treasures of nature, and had ever been victorious.

Asa Gray, then in his early twenties, had visited Albany a few years earlier, and had first seen Henry, with whom he was to have a long and firm friendship. Gray found him a "grave-looking man" and let that brief description stand unadorned.[4]

But the personal appearance of the new professor and the originality of his laboratory attainments were not the principal questions to which faculty and students of Princeton awaited a reply. The great question was: Would he prove to be a good teacher?

Henry had given proof of being a clear expositor (although his

[4] *Letters of Asa Gray.* Vol. I, p. 15.

clarity at no time suggested a lack of depth) and of possessing a gift for apt illustration. He had gained some advanced views upon educational methods during the course of his friendship with Amos Eaton and had eagerly embraced the Lancaster system of education, although its inventor was described by Sir Walter Scott as a mountebank. But he had to readjust his views and to plant them on a broader basis.

His course with the technical students was tolerably clear. With the general student the way was not so clear, for he had to make this student understand the value of natural science as a mental discipline, as an instrument for training the faculties of observation and reasoning. Only a very few of his students would follow careers in which they would undertake chemical or physical investigations, but he held the broad view of trying to persuade everyone of them that they should be interested in the acquisition of truth and its relations, whatever their work might be. By understanding how truth was won and knowledge advanced in one field of inquiry, they would gain an aptitude in any form of investigation.

He exercised his students in deducing particular laws from general laws, and then in ascertaining whether what had been deduced actually occurred in nature. One of his students recorded the impression he made: "No one who was under his instruction can ever forget his definition of science, or his manner of enunciating it with his handsome face and magnificent physique. 'Science, gentlemen, is the knowledge of the laws of phenomena, whether they relate to mind or to matter.' " [5]

Henry did not put his views on education into print until 1854, after he had delivered them in the form of an introductory address as the retiring president of the Association for the Advancement of Education.[6] At that date they had been sharpened by his experience at Princeton and by association with Alexander Dallas Bache, Benjamin Franklin's great-grandson, first president of the unique Girard College in Philadelphia, and professor at the University of Pennsylvania. Henry disclaimed any book knowledge on the subject but what he lacked in reading was more than compensated by experience and by discussions with this brilliant friend, who could have introduced him to Victor Cousin's *Report on Public Instruction in Prussia,* which

---

[5] Cameron, H. C. Reminiscences. *Memorial,* p. 170.
[6] *American Journal of Education.* Vol. I, 1855, p. 17; *S.W.* Vol. I, p. 325.

had a wide circulation in this country and had been studied closely by Bache.

At the very outset he laid emphasis upon education as a form of discipline, an attitude which might disturb some modern educators and parents, but Henry was aware that much of what he had to say was at variance with the established principles even in his own day. "The first remark which may be made in regard to education is that it is a forced condition of mind and body. As a general rule it is produced by coercion—at the expense of labor on the part of the educator, and of toil and effort on the part of the instructed." The student may be encouraged by the prospect of reward and by appeals to the affection but, when these prove ineffectual, the teacher should be empowered to apply the rod, since pain should be associated with wrongdoing. Some modern educators will dispute his assertion that the child "if left to himself, would receive no proper development, although surrounded with influences which would materially affect its condition. The savage never educates himself mentally."

Henry perceived a great disparity in the rewards to those who receive a higher education. "In order that civilization may continue to advance, it therefore becomes necessary that special provision should be made for the *actual increase* of knowledge, as well as its diffusion; and that support should be afforded, rewards given, and honors conferred, on those who really add to the sum of knowledge. This truth, however, is not generally appreciated and the tendency is to look merely at the results of the application of science to art, and to liberally endow and honor those who simply apply known facts rather than those who *discover* new principles."

Although these views were expressed nearly a hundred years ago they are reflected in the words of those educators of today who are alarmed by the diversion of young scientists from the university and research institutes into industrial laboratories where the salaries are higher. In the general diffusion of elementary education without a proper corrective in the form of a cultivation of the higher intellectual qualities, Henry could foresee nothing but the danger of promoting an inordinate love of wealth.

He might have been speaking to a modern audience when he said of the tendency of the public education of his day: "In this country, so far as I have observed, the course of education is defective in two extremes: it is defective in not imparting the mental habits or facilities which can most easily be acquired in early life, and it is equally defective in the other extreme, in not instructing the student, at the proper period, in processes of logical thought, or deductions from general principles. While elementary schools profess to teach almost the whole circle of knowledge, and neglect to impart those essential processes of mental art of which we have before spoken, our higher institutions, with some exceptions, fail to impart knowledge, except that which is of a superficial nature. The value of facts, rather than of principles, is inculcated. The one however is almost a consequence of the other. If proper seeds are not sown, a valuable harvest cannot be reaped."

A high state of civilization could be preserved only by the exertions of individuals actuated by "generous, liberal, and enlightened philanthropy." Unfortunately, the increase of ease and security tended to relax individual effort. "Man is naturally an indolent being, and unless actuated by strong inducements or educated by coercion to habits of industry, his tendency is toward supineness and inaction." The growth of wealth and elementary education without a corresponding increase of higher education rendered the voice of the serious teacher less and less audible, and obliged him to comply with popular prejudices.

The growth of specialization in research, he foresaw, would narrow the number of persons thoroughly conversant with any field of knowledge. These small groups of well-informed persons exercised only small influence and were generally not the dispensers of favor. He who aspired to wealth did not seek their approbation. Those men who are actively engaged in the business of life have little time for profound thought. They are condemned to receive their knowledge at second hand and yet "they are not content under our present system of education with the position of students." They aspire to be regarded as teachers, and are ambitious to become authors, impatient for applause. "In such a condition of things it is possible that the directing power of an age may become less and less intelligent as it becomes more authoritative, and that the world may be actually declining in what constitutes real moral and intellectual greatness, while to the superficial observer it appears to be in a state of rapid advance."

He lamented that while his was emphatically a reading age it was not proportionately a thinking age. The world suffered from too many bad books and too much ill-digested reading. He laid great

Philosophical Hall, which housed Henry's laboratory in Princeton.

Joseph Henry's home on the Princeton campus. In the background is Stanhope Hall, the twin of Philosophical Hall. This picture was taken shortly before Joseph Henry House was moved in 1870.

stress upon the early training in habits because persons trained in good habits will act well in times of crisis where they have no time to think.

Henry was careful to point out that he was criticizing not the existing state of education but rather the tendencies, and it will be conceded that he had a clear perception of the outcome. So much of what he foresaw came about as he predicted.

Henry gave too much stress to the correction of deficiencies and not enough to the cultivation of aptitudes, to the building up of interest and capacity. However, that is the function of the teacher. Henry knew that intellect cannot be created artificially, nor can originality be taught, but in all his teaching at Princeton it was his purpose to see that what intellect and originality existed should be directed into fertile channels, so that those who had the gift of connecting facts should not fail because the facts were not available.

Although these thoughts on education were of a riper year, the principles he expounded must have been more or less apparent to him when he assumed his duties at Princeton.

Was Henry a good teacher?

It is difficult to define "a good teacher." The views of students point toward a popular teacher, while those of the teaching profession tend to overlook personality. Henry brought to his work at Princeton certain qualities without which no teacher is successful. He had patience and enthusiasm, a desire to make his students want to learn. No one could lecture as Henry did without being interested in his audience. He was less concerned with being a teacher of science than he was with giving his students new interests, wider horizons, and sound habits of thinking.

That he was eminently successful in attaining these objects is testified by the old students who were called upon to perform the melancholy duty of delivering memorial addresses at the time of his death. It will be observed that few of these men became professional scientists, but they all recalled with joy and thankfulness the association with their old professor.

It was no light matter to fill a chair in a long established college with credit, but Henry must have been encouraged by the attitude of his faculty colleagues. They do not appear to have entertained any doubts about his abilities as a teacher. Dr. Torrey went off to visit Europe in the first year of Henry's residence, but he asked the newcomer to take his classes in chemistry, geology, and mineralogy during his absence. Torrey was already acquainted with Henry through their mutual friendship with Eaton, and while the pair were together at Princeton they visited each other daily. Torrey's confidence in Henry's ability to undertake the additional duties must have helped the other members of the staff to regard their new colleague with favor, for he substituted in other chairs when the occupants were absent.

He lectured in astronomy, a subject in which he undoubtedly had improved his knowledge through association with Stephen Alexander, who had accompanied him to Princeton where he was entered as a student of theology. However, it is not so easy to understand how he came to be asked to lecture on architecture.

The busy first year at Princeton, when he had to fill two chairs, did not permit much leisure for original research. His lecture room was in the upper story of Philosophical Hall, since demolished. The apparatus was kept in glass cases which covered the walls, while the students' seats occupied the floor of the room. A card, probably obtained from a local tradesman, displayed the legend "A place for everything and everything in its place," thus admonishing the students that orderliness was expected of them and, as the legend was embellished with the picture of a whip, a stern attitude against transgressors was implied.

Among the apparatus brought by Henry from Albany and given places of honor in the glass covered cases were the little magnet with which his earliest experiments had been performed, and the machine with which he had first demonstrated how mechanical motion could be produced by electromagnetic impulses. These were the tools on which his fame as a scientist were based, but they were likewise mementos of his early struggles when materials were harder to obtain.

One of the advantages offered by the transfer to Princeton was a greater ease in obtaining materials. Not that the college was richly endowed, for the era had not yet dawned when research funds were readily furnished. But the college could provide him with some assistance for the construction of apparatus, and he had a larger salary upon which to draw when he wished to supplement what the college provided. In addition, he was near the centers of industry. Proximity to New York and Philadelphia greatly facilitated the acquisition of materials.

When time permitted he undertook the construction of a large magnet, one capable of sustaining 2,500 pounds. This was the largest magnet he had possessed for his own use. The making of it occupied most of his leisure during that first crowded year. To activate the magnet he constructed an original galvanic battery. This battery was intended to be highly flexible since it was intended to exhibit most galvanic and electromagnetic phenomena on a large scale without recourse to any other battery. The several parts of the battery were not soldered together, but the design was such that the units could be connected temporarily by means of movable conductors, so arranged that any number of elements, from a single pair to eighty-eight pairs could be brought into action. Although all the plates could be used together as a single group, the battery was divided into eight sub-batteries of eleven pairs each, and these could be used in any desired combination by means of a crank and a windlass shaft.

After reading the descriptive details of this remarkable battery [7] one may well believe that its arrangements were "adopted in most cases after several experiments and much personal labor." It was a most ingenious contrivance.

Splendid as the magnet appeared to be when it was completed, it did not satisfy Henry. The inventor of the powerful electromagnet could not rest satisfied until he had the most powerful specimen in the country. He had to have one which would far surpass the famous Yale magnet. Accordingly, he began the construction of a second. This one sustained a weight of 3,500 pounds. He was delighted with it. One of his students said of this magnet: "We shall always remember the intense eagerness with which he superintended and watched the preparations, and how he fairly leaped from the floor in excitement when he saw his instrument suspending and holding a weight of more than a ton and a half." [8]

These magnets were not designed to provide the professor with the barren satisfaction of possessing an instrument more powerful than that owned by anyone else. They were designed to permit the continuation of his researches into electromagnetism with the best apparatus available. He had had the idea in mind while he was at Albany, now he was realizing it.

The investigations Henry conducted at Princeton constitute some

[7] Am. Phil. Soc. *Trans.* Vol. V (n.s.) 1837, p. 217; *S.W.* Vol. I, p. 80.
[8] Dod, S. B. Discourse. *Memorial*, p. 141.

PRINCETON

of the most acute and penetrating observations of the rapidly expanding science of electricity. Their value does not reside in the discovery or the observation at the stage where Henry left it, so much as in the developments made by other men who took up the work at the point where Henry left it. Henry was the prophet who gave inspiration to a new school of science in America. Through these experiments he became a leader who opened new territories for investigation by other thinkers.

Another advantage offered by removal to Princeton was the introduction it afforded to a wider circle of men engaged in scientific pursuits. Communion with men who, like himself, were engaged in the advancement of knowledge had been largely denied him while he was in Albany. Now he was no longer obliged to work in an intellectual vacuum. Not only was his society reinforced by the larger companionships of the men who formed the college faculty, but he was within easy reach of other active centers of intelligent men, interested in technical subjects.

A short journey on the newly opened railroad carried him to Philadelphia, seat of another university and home of several learned societies, notably the American Philosophical Society and The Franklin Institute. Henry gravitated to this city rather than to New York because of the attraction of his devoted friend Alexander Dallas Bache, who was an active member of both the societies mentioned.

Henry did not display a conspicuous part in The Franklin Institute, although he was a member of its Committee on Science and the Arts, a group which concerned itself with securing recognition for inventions of merit. A resolution passed at the time of his death alludes to his contributions to the Institute's *Journal* but, with one exception, none of these can be identified.

He was soon at home with the members of the American Philosophical Society, to which he was elected in 1835, and it was to the members of this society that he communicated the results of his Princeton researches. Probably through the instrumentality of Bache, Henry made the acquaintance of Robert Hare, professor of chemistry at the Medical School of the University of Pennsylvania. Hare was already distinguished for the discovery of the compound blowpipe and other contributions to chemistry, but he was also an enthusiastic investigator of electrical phenomena, and had built two instruments, the calorimotor and the deflagrator, which added to his distinction. Hare and Henry struck up a friendship and were soon associated in the performance of electrical experiments, in which Henry was the manipulator.[9]

By 1835, when no telegraph wires existed elsewhere, he had strung his own wires in front and rear of Nassau Hall. The erection of wires across the campus aroused a lively curiosity among the students. Glancing at the strange wires stretched beneath the elms, they would not have been normal college students had they refrained from speculating at their purpose, or abstained from pleasantries when they learned that the purpose of the wires was to enable the professor to communicate with his wife.

Evidently Henry perceived some of the potentialities of the primitive telegraph he had constructed in Albany if he stretched a line over several hundred feet of ground so soon after his arrival at Princeton. The new telegraph was an improvement upon its predecessor. The current in the circuit excited a small magnet which pulled toward it a short iron bar laid across the poles and carried on one end of a wooden lever, on the other end of which was a hammer by which a bell was struck.[10] But the principal innovation was that the circuit was completed through the earth by sinking the ends of the wire in two wells.

Henry was not the originator of the ground return, as is sometimes asserted. The grounded circuit had already been used in electrostatic telegraphs and in firing gunpowder at a distance by Cavallo in 1794 and Salva in 1798.[11]

However, this period is marked by another discovery of importance. This was Henry's construction of the electromagnet relay, an essential feature of the electrical telegraph which came into universal use.

With an intensity magnet in the circuit of which we have just spoken, he was able to move a forked wire dipping into two mercury cups so as to make or to break the local circuit of his large quantity magnet.

Unhappily, the experience of delaying the announcement of the discovery of mutual induction had not taught him how important it was to publish the accounts of his work if he were to be credited

---

[9] Smith, E. F. *Life of Robert Hare*, p. 187.
[10] Magie. *Op. cit.*, p. 475.
[11] Fahie, J. J. *History of the Electric Telegraph to the year 1837*. London, 1884, pp. 99, 108.

with the original idea. Although there were no reports published at the time, the idea was exhibited to Henry's students and friends. Thus Henry's claim, advanced at a later date, was corroborated by the statements of others. Edward N. Dickerson, a Princeton student of 1839, declared that the relay was put into operation in 1835.[12] Henry's description of the relay given in his deposition in the case of *Morse v. O'Reilly,* is as follows:

> In February of 1837 I went to Europe; and early in April of that year Professor Wheatstone, of London, in the course of a visit to him in King's College, London, with Prof. Bache, now of the Coast Survey, explained to us his plans of an electro-magnetic telegraph; and among other things exhibited to us his method of bringing into action a second galvanic current. . . . I informed him that I had devised another method of producing effects somewhat similar. This consisted in opening the circuit of my large quantity magnet at Princeton, when loaded with many hundred pounds weight, by attracting upward a small piece of moveable wire, with a small intensity magnet connected with a long wire circuit. When the current of the large battery was thus broken by an action from a distance, the weights would fall, and great mechanical effects could thus be produced, such as the ringing of church bells at a distance of a hundred miles or more, an illustration of which I had previously given my class at Princeton. . . . The object of Professor Wheatstone, as I understood it, in bringing into action a second circuit, was to provide a remedy for the diminution of force in a long circuit. My object, in the process described by me, was to bring into operation a large quantity magnet, connected with a quantity battery in a local circuit, by means of a small intensity magnet, and an intensity battery at a distance.[13]

This invention is, of course, fundamental to the operation of electrical systems of many kinds—telephone, telegraph, and electric power.

[12] Dickerson, E. N. *Joseph Henry and the Magnetic Telegraph: an address,* 1885.
[13] Smith. Inst. *Annual Report for 1857,* p. 111.

Henry first addressed the American Philosophical Society on January 16, 1835. This address was the first of a series which were grouped under the general title of "Contributions to Electricity." In the address he described the elaborate crank-operated galvanic battery for producing electricities of different intensities. From the description it can be seen he was employing it in essentially the same manner that storage battery installations are used today for charging them in parallel and discharging them in series combinations for high voltage service. The principle was the same, although the nomenclature differed.

This was not the only communication of the session. He gave an extemporaneous talk to the members which was published under the title of "Facts in reference to the Spark, etc. from a Long Conductor uniting the Poles of a Galvanic Battery." Within a few days, there arrived in this country the November (1834) issue of the *London and Edinburgh Journal of Science* containing an article by Faraday entitled "On a peculiar condition of Electric and Magneto-electric Induction." In this paper the English scientist described the phenomenon of self-induction which Henry had announced two and a half years earlier. Faraday was evidently not interested in what American scientists were doing, and did not read Silliman's *Journal* in which Henry had made the disclosure.

This oversight is all the more curious when one recalls that Faraday had only succeeded in his electrical investigations after he had adopted Henry's magnets. One would have thought that he would have been keenly interested in the work of the American scientist who had given proof of an ability to make outstanding contributions to the science of electricity.

When the news of Faraday's "discovery" of self-induction reached Philadelphia it incited Bache to persuade Henry that no time should be lost in publishing his remarks before the members of the American Philosophical Society. So little was then known about electricity that an important principle might lie unperceived in almost any experiment until an observant eye detected it in reading the account describing it. Bache was resolved that his friend Henry should not be robbed of any further priorities in publication, but the obstacle in the way of obtaining prompt publicity for Henry's latest contribution to self-induction lay in the Society's arrangements for the publication of its papers. The Society was as leisurely in its publishing practices as was Henry himself. The paper he had delivered in January 1835 would not appear in print in the *Transactions* until sometime in 1837.

Bache was a member of the editorial board for the Franklin Institute's *Journal*. He decided that this should be the vehicle for conveying Henry's latest thoughts on self-induction to the world. At least immediate publication would guarantee priority for any original observation contained in the paper. At Bache's request an abstract of Henry's remarks was published as "an authorized extract from the proceedings of the stated meeting of the American Philosophical Society." [14]

Bache's solicitous action ensured a brief priority for Henry in the discovery of non-inductive windings, the same in principle for those now used for resistance coils and other apparatus where self-inductance is reduced to a minimum. This occurs in the paragraph which reads:

"6. A ribbon of copper, first doubled into two strands and then coiled in a flat spiral, gives no spark, or a very feeble one."

The importance of this brief announcement became significant three weeks later when Henry had an opportunity to elaborate upon it. On February 6, 1835, he presented the second paper in the series of "Contributions to Electricity," which embraced the experiments in self-induction he had performed during the previous year. The motive for publication is made clear in the introduction. The experiments, "though not as complete as I would wish, are now presented to the Society, with the belief that they will be interesting at this time on account of the recent publication of Mr. Faraday on the same subject."

Henry quotes his account of the discovery of self-induction given in Silliman's *Journal* [15] and then proceeds to demonstrate how far he had progressed in the subject during the interval. For the better comprehension of his words it should be noted that he was now employing a new form of conductor. This consisted of a ribbon or strip of copper nearly one inch in width and 28 feet long, insulated with silk, and coiled into a spiral like a clock spring. This proved so satisfactory that he sought longer similar ribbons and eventually obtained one 96 feet long.

[14] *Journal of the Franklin Institute*. Vol. XV, 1835, p. 169; *S.W.* Vol. I, p. 87.
[15] See *ante* p. 89.

PRINCETON

The elaboration of the discovery of non-inductive windings appears in the paragraph which reads:

> One of these ribbons was next doubled into two equal strands, and then rolled into a double spiral with the point of doubling at the centre. By this arrangement the electricity, in passing through the spiral, would move in opposite directions in each contiguous spire, and it was supposed that in this case the opposite actions which might be produced would neutralize each other. The result was in accordance with the anticipation; the double spiral gave no spark whatever, while the other ribbon coiled into a single spiral produced as before a loud snap. Lest the effect might be due to some accidental touching of the different spires, the double spiral was covered with an additional coating of silk, and also the other ribbon was coiled in the same manner, the effect with both was the same.[16]

It is interesting to observe that prior to this, in 1832, Faraday had constructed for use in his experiments on mutual induction an arrangement closely similar to the non-inductive winding. This consisted of two secondary windings twisted about each other to associate them as closely as possible, and then wound into the form of a helix. With these windings connected in series opposing, he found no mutual inductive effect.[17] But Faraday's first published reference to non-inductive windings occurred in the paper he read before the Royal Society, January 29, 1835, and published in the *Philosophical Transactions* of that year. He described the non-inductive windings as follows:

> Thus, if a long wire be doubled so that the current in the two halves shall have opposite actions, it ought not to give a sensible spark at the moment of disjunction; and this proved to be the case, for a wire forty feet long, covered with silk, being doubled and tied closely together within four inches of the extremities, when used in that state, gave scarcely a perceptible spark.[18]

[16] *S.W.* Vol. I, p. 94.
[17] Faraday. *Exper. Researches.* Vol. I, para. 194, p. 57.
[18] Roy. Soc. *Phil. Trans.* 1835, part I, p. 50: *Exper. Researches.* Vol. I, para. 1096, p. 335.

We perceive the two men still working closely along parallel paths and again reaching similar results at about the same time.

In the same paper, Henry described a number of other experiments to investigate the phenomenon of self-induction which led him to infer "that some effects hitherto attributed to magneto-electric action are chiefly due to the reaction on each other of the several spires of the coil which surround the magnet." In additional experiments it was observed that the sparks drawn from the spirals resembled the discharges from a Leyden jar.

The paper reveals clearly the handicaps under which Henry and his contemporaries labored in having no reliable or sensitive meters. The pioneers in electrostatics had literally to "feel" their way by physical shocks. The measurement of an electrical charge by the intensity of the physical shock it produced was far from precise, but it provided some qualitative estimate. Another method, with pith balls singly or in pairs, was more informative. The measurement of the angle of separation of a gold-leaf electroscope to determine the amount of electrical charge, introduced in 1786, was an advance in method, but difficult to apply in the case of transient charges. Henry had found that fixed coil galvanometers, which had become available, were not sufficiently sensitive for measuring the transient currents in secondary circuits, so he reverted to the most primitive measurement of all, the physical shock.

In this paper he tells of detecting the presence of a small current by placing the end of the conductor in his mouth. More intense charges were measured by the sensations experienced in his fingers, wrists, or arms.

Before Henry made any further reports on his investigations into self-induction he was granted the opportunity of meeting Faraday, for whom he had conceived a high admiration. The meeting was rendered possible when, in 1836, the trustees of the college granted Henry a year's leave of absence on full salary in recognition of his work during his first four years at Princeton. It was a good time to escape from the United States, for the nation was just entering upon one of those financial depressions which recur periodically to remind us that we are not yet the masters of our economic system.

Arrangements were made for the journey in company with Richard Varick De Witt, his old friend of the Albany Institute, and by a lucky chance his new friend Bache and his charming young wife were also sailing for Europe at about the same time. Bache was visiting Europe to study the educational systems of that continent on behalf of the Stephen Girard Foundation in anticipation of the opening of Girard College, to which he had been appointed principal.

Henry's preparations for the journey were begun in November 1836, when he went with De Witt to New York. In the following month they set out for Washington, where they proposed to obtain letters of introduction.

First impressions of the national capital were not favorable. Writing to Harriet, Henry complained that, although they had taken lodgings "at the principal house, Gaisby's Hotel," they did not find the accommodation as good as houses of the same class in Philadelphia or New York. Nor did he form a good impression of the Capitol, the effect of which "at first sight and from a distance is not good. The pillars of the Corinthian or Composite order are too slender to appear well at a distance; when, however, the eye is brought sufficiently near to take in the details of the structure the effect is very imposing. The style of architecture is that called Italo-Romanesque, which was very much in vogue before the study of the Grecian remains introduced a better taste. As far as the Architecture is concerned it would scarcely be a loss were the British again to burn the Capitol—in that case I am sure, a more imposing and, at the same time, more simple building would be erected in its stead."

His letters to Harriet are full of instruction, too full. For during this visit to Washington he met some of the most interesting political figures of the time, but he had very little to say about them. He heard John Quincy Adams address Congress, and met him later in the day. "He is to my eye totally unlike the portraits of him. He has very little hair, and this is not white but sprinkled with black." He was much pleased with Senator John C. Calhoun of South Carolina, finding him a man of intelligence, interested in the cause of science. They were to become close friends later in life. Henry also called upon the newly elected president, Martin Van Buren, who received him cordially and promised letters of introduction.

Meanwhile Torrey had been commissioned to book passages for the travellers. Early in February he reported: "I have taken state rooms for two passengers in the splendid ship *Wellington* [19] which sails for London on the 20th inst. She is a noble vessel, entirely new, having

[19] The "splendid" and "noble" vessel was a modest packet of 750 tons.

never yet made a voyage, so you will not be annoyed by bilge water nor by the creaking of masts and joints, the noise of which is so irksome in stormy weather. The state rooms are large and all the accommodations are excellent. . . . Captain Chadwick is a fine fellow, he says he hopes to reach Portsmouth in eighteen days."

When the news of their professor's impending departure was communicated to the members of the Princeton senior class they appointed a committee of three students "to express their kind feelings toward Professor Henry" and to wish him well on his journey.

The voyage was uneventful. In his first letter to Harriet, Henry wrote: "Although when the sea was not rough I was tolerably comfortable, I could fix my mind on nothing and the whole time passed on the water appears less than the week I have passed in London. Our ship's company was very pleasant and we had every attention and luxury which could be desired, but I was very glad to escape from the dominion of Neptune and happy again to be on Terra Firma."

Captain Chadwick's confidence in his new vessel's ability to make Portsmouth in eighteen days might not have been misplaced but for an easterly wind which smote them just as they came off the English shore. This wind blew with such violence that the restless passengers began to despair of ever being able to beat up the English Channel. When the wind moderated somewhat, the Captain offered to put a boat ashore at Plymouth for the more impatient passengers who were willing to pay for the journey overland to London. Henry and De Witt were among those who chose to land.

It was not impatience so much as curiosity which prompted Henry to land at Plymouth, since this permitted him to call upon Sir William Snow Harris at the Royal Dockyard. Sir William had gained a reputation for his experiments in electrical attraction and repulsion, in which he had employed some unusually ingenious apparatus. Henry looked forward to meeting any English scientist who was working in this field, so the opportunity of visiting Harris was not to be missed.

In one respect it was a fortunate prelude to the English visit, for his welcome set him in the right frame of mind to enjoy the visit. Sir William Harris proved to be an exceptional host. Not only did he admit Henry to his laboratory and display all his apparatus but he went further and willingly outlined his projected course of study. The host insisted that his American guests spend the entire day with him.

PRINCETON

The journal of the European visit is in many respects a disappointing document. Other American visitors, notably Silliman, had kept full diaries of their journey which, when published, had been greedily devoured by the public at home. Henry probably never contemplated publishing his journal; it is more likely that he wrote it as a personal memento of the men he met, retaining in his memory the more interesting details of their association.

The visitor to England at the time could have perceived much that would have interested an alert American social observer. It was in many respects a new England compared with that pictured by the American nurtured upon ideas of colonial tyranny. After it appeared that a social revolution was inevitable, the nation had emerged into a clearer atmosphere rendered possible by a reform in political thought. The spirit of industry had taken a firm grip upon the people, and with the advent of the railroad a new era of progress had arrived.

Had Henry been more interested in man and mankind he would have found much to engage his attention, but he never was deeply interested in the human problems and his mind was not awakened to the profound influence scientific investigations must have upon human progress during the coming generation. The ferment of ideas had not yet penetrated the social circles to which he carried introductions. The Oxford dons, the Eton masters, and their kindred regarded the future with apprehension, so that it is not improbable this feeling permeated the thinking of scientists who turned their questioning glances toward the society in which their days were cast.

It is a matter of regret that Henry did not come into contact with some acute politically minded person who could have pointed out to him the transition from the policy of *laissez faire* to one that was concerned with Factory Acts. Had he become aware of the great movement that was stirring in England, he would have been prepared to foresee the events that must spring from a similar change in America. The vision of a new relationship between science and industry escaped him. This is to be noted not so much as a defect in Henry, for he was not alone in his blindness; but in view of the leadership he was eventually to assume, a foresight of the future drawn from the British example would have enabled him to wield a beneficial influence in this sphere of action.

If he went to England unprepared to bear his share in the table talk upon Factory Acts, company promotion, and race horses, he had a pulse ready to be stirred by the love of knowledge and the joy of scientific discovery. His journal is devoted to what he could learn from fellow scientists, and when none was at hand to gratify his hunger in this respect, the daily record is filled with a dreary account of the tourist. For example, on the first day of his arrival, having learned all he could about his host's electrical studies, he visited the dockyard and dutifully records its acreage, the number of employees, the length of the masts in storage, and other unilluminating details. He anticipated meeting the famous engineer Isambard Brunel, but he failed to observe Brunel's installation of the division of labor in the making of wooden blocks and pulleys which had only recently been introduced into the British naval dockyards and which was, without doubt, the most important feature on display. It was the first step of its kind, and its only American parallel was probably to be found in Eli Whitney's gunshop at New Haven.

But if Henry missed the first signs of the industrial future it must not be supposed that he was a completely unimaginative traveller. There are glimpses of a mind susceptible to other manners and ideas. His journey by coach from Plymouth to London provided an incident which became deeply imbedded in his mind.

Learning that the coach would traverse Salisbury Plain in the dead of night, Henry tipped the driver to make a short halt which would enable him to inspect the ruins of Stonehenge. Anyone who has visited that famous prehistoric monument by daytime would have found the scene commonplace enough, but Henry chanced to see the array of monoliths under the light of a bright moon, and one might almost believe he communed with the ghosts of the place. Long afterwards he recalled the eeriness of the scene and the awe which assailed him as he scrambled alone among the relics of the Druidical age.

He reached London on March 17, put up temporarily at the "Bull and Mouth," an undistinguished tavern, and spent the next day looking for lodgings. These he found at 37 Jermyn Street, "in a pleasant part of the city within a few steps of the Palace of St. James and about the same distance from the Royal Institution. I have often, as you may suppose, visited the latter but do not intend to pay much court to the former. Queen Adelaide is not interesting to me."

Having found a *pied à terre* he presented his letter from the President-elect to Mr. Stevenson, the ambassador from Washington, from whom he received a cordial reception and the offer of help should he need it in meeting people.

As the time of the arrival of the visitors happened to be Easter Week, they were disappointed to learn that even the scientists were in holiday mood and had moved in a body from the city. The visitors thereon went in pursuit of their fellow-countrymen who, they felt, would not succumb to such weakness. Nor were they disappointed. Another introductory letter which Henry bore was the first link in a chain of circumstances which was to control Henry's later life.

In his association with the members of the Franklin Institute in Philadelphia, he had formed an intimate acquaintance with Henry Vaughan, who gave him a letter to his brother, then resident in London. At the first dinner in Vaughan's home a fellow guest proved to be Richard Rush, former ambassador to Great Britain, who was then in London as the representative of the American government in the settlement of Hugh Smithson's will. Thus, at this very early stage, Henry was made acquainted with the singular bequest for the foundation of the Smithsonian Institution, with which the second half of his life was to be intimately interwoven.

Henry lost no time in calling at the offices of the Royal Society and the Royal Institution. By mischance, Faraday had just left on his Easter vacation. This circumstance, coupled with the arrival of Professor and Mrs. Bache in London, furnished him with a brief respite from science (although he could not be prevented from hunting down the lesser lights, whose importance grew because of Faraday's absence) and permitted him to see some of the city's sights. London was not as he had pictured it. "To my surprise every building in London is as black as if formed of cast iron. The terraces which look so well on paper in reality are as if powdered with a thick coat of lampblack, which is washed off in places and then variegated with irregular stripes of white. The effect is such as to interfere with the pleasure in the building."

He was frank in his admiration for the beauty of the interior of St. Paul's cathedral, although his mind rebelled against the recognition of fame as demonstrated by its monuments to the illustrious dead. "The inside of St. Paul's is magnificent . . . the walls are studded in almost every part with statues and monuments to the departed greatness of the English people. It is, however, a commentary upon the character of the last two centuries of this people that there are but three or four monuments to men of literature and science, while the walls are covered with monuments of military and naval heroes. It is the same in Westminster Abbey. While the poet occupies a small corner and is commemorated by a small stone, the hero of barbarous war lies in a marble tomb reaching in many cases almost to the vaulted roof, and many names unknown to general fame live only in this manner, immortalized by friends or circumstances."

Reflections like this receive expression only in the letters to his wife and to Stephen Alexander. The bulk of the letters and the whole of the journal are devoted to severely practical matters. Some time had to be devoted to the search for apparatus. In those days there were few pieces of familiar and proved apparatus which could be chosen from manufacturers' catalogues and which form the stock-in-trade of scientific lecturers of modern times. As Henry visited laboratories and saw apparatus which aroused his admiration he lost no time in calling upon the makers to have duplicates made. His journal records numerous conferences with instrument makers.

His letters are sprinkled with the names of the great and the near great in science—Faraday, Wheatstone, Daniell, Sabine, Sturgeon, Babbage, and many others. It must have been flattering to find that everywhere he went he met with the friendliest reception. The English scientists received him with the honor that was his due.

Naturally, Faraday was the greatest attraction, and the pair contrived to spend some time together in spite of Faraday's stern scheduling of the time he devoted to intercourse with his fellow-beings. The Englishman had formed such a high opinion of his visitor's achievements that he almost succeeded in persuading him to address the members of the Royal Institution on the mathematical aspect of his work. Most unfortunately, he failed to influence Henry. The latter modestly repulsed his friend's approaches with the comment that he had come to learn and not to instruct. Thus, a valuable and much needed paper on Henry's work was omitted from the record.

Henry attended Faraday's course of lectures on metals. "The subject is one of the dullest in the whole course of chemistry, but by his manner and the many new views he gives, the lectures are interesting and very instructive." Having heard Faraday give one of his popular lectures, to an audience composed largely of ladies, on the subject of the four elements of the ancients, he made the remark: "Professor Faraday is deservedly a very popular lecturer, but does not surprise or strike one with the depth of his remarks, or the power of a profound mind, but moves by his vivacity of manner and his happy illustrations, as well as his inimitable tact in experimenting."

When Faraday's exacting schedule caused him to deny himself to his visitor, Henry got along very well with Mrs. Faraday, who took him along on a round of visits. He has this comment to make in the journal: "Mrs. Faraday is unassuming and well read, of not much personal pretensions, yet with an agreeable countenance. She gave us some anecdotes about her husband; and said he was about forty-five years old—fond of novels—not disposed to go into company, but associates principally with a few persons—denies himself to everyone three days in the week—never dines out, except when commanded by the Duke of Sussex at the anniversary of the Royal Institution.

"Mrs. Faraday seldom goes out and sees but few persons. She says she is a good nurse, and that Mr. Faraday puts himself implicitly under her directions. She prescribed for my cold."

If one had no other guide but this one might be persuaded to believe that Faraday was something of an old curmudgeon, but Henry does not generalize about people any more than he did about science. During the part of the week when he was excluded from Faraday, Henry spent most of his time with Charles Wheatstone, professor of experimental philosophy at King's College, London. With him Henry spent many delightful hours discussing their common interest, the electric telegraph. On one occasion Faraday, Daniell (of battery fame), and Wheatstone (whom any Englishman will tell you invented the telegraph) joined in making experiments in electricity in the latter's laboratory. One of the experiments related to the extraction of a spark from current generated by a thermopile. All three Englishmen essayed the experiment without success, and were disposed to abandon it as impracticable. Henry asked permission to make the attempt. He took the precaution of increasing the self-induction of the current by sending it through a long wire wrapped around a piece of soft iron. This produced the desired spark.

Faraday demonstrated his delight by jumping to his feet with the joyful exclamation, "Hurrah for the Yankee experiment!" The three Englishmen recognized that they were in the presence of the man who had discovered self-induction and who knew more about it than anyone else.

King's College still preserves two coils of copper strip insulated with silk which, the descriptive label says, were used by Henry on this visit, and tradition declares they were used in the experiment conducted in the presence of Faraday, Daniell, and Wheatstone.[20]

At the time of Henry's visit, Wheatstone was deep in his experiments with the needle-telegraph which he invented. Delighted to have an intelligent listener, he freely demonstrated his work. Like Morse, in America, the English professor was wrestling unsuccessfully with the difficulties of transmitting his signals over a long wire. Wheatstone demonstrated his arrangement of a supplementary local circuit from an additional battery. Henry explained his own much simpler relay. The Englishman later admitted that Henry showed him how to use a galvanometer for detecting a current at the end of a long conductor, but he never publicly acknowledged his indebtedness to Henry for information about the relay. The nearest he ever came to acknowledgement was the concession that he only claimed for himself the credit of being the first to use it *in England*.

However, it would be doing Wheatstone an injustice to attribute to him a want of appreciation for Henry's achievements. A few years later he joined Faraday in recommending the Royal Society to award Henry the Copley Medal, which had been given to one American only, Benjamin Franklin (we exclude Rumford, whose work had been performed in Europe). The proposal was deferred by the Council under pressure of other business and thereafter it was overlooked. This valued award of scientists to a scientist has since been awarded to men less deserving of the honor.

The last few days of Henry's stay in England were devoted to writing an analysis of the experiments and reports of The Franklin Institute committee on the bursting of steam boilers. Since the use of high-pressure steam had made more rapid progress in America under the influence of Oliver Evans than in England, the information Henry had to furnish was eagerly awaited by British engineers. The Institute tests were historic, and the editor who asked for an account of them, with the findings of the committee, was only yielding to the desire of practical men to learn more about enlarged power, the use of which had been retarded by Watt's conviction that boiler construction was not strong enough to withstand the higher pressures. Henry had attended some of the investigating committee's meetings, but

---

[20] A photograph of the coils is reproduced in Appleyard, R. *Pioneers in Electrical Communication.* London, 1930, p. 104.

he undertook no light task when he consented to write a summary of their work. In one of his letters home, he admitted that the writing of the article "cost me several days of assiduous labor." Even so, the article does not appear to have been printed.

Eight weeks had passed since his arrival in England and, while he confessed that he had not exhausted the wonders of science and art in Britain, he felt "I have obtained as much information as will, I hope, pay for the time and money expended. . . . I feel much regret in parting from the kind friends I have made in this place."

He arrived in Paris on May 10 and took a room in the Hotel Normandie, on the Rue Saint Honoré. For two weeks he did little more than tour the city, enjoying the strange sights and wondering at the ways of the people. There was little to be gained by visiting the men whose acquaintance he courted since his limited knowledge of the spoken language prevented him from exchanging ideas with French scholars. By good fortune he met Thomas Hun, a friend of Albany days, who was residing in the Latin Quarter while studying medicine. Hun's services as an interpreter were promptly enlisted. In later years, when he was a well-known medical practitioner, Hun used to boast facetiously that he had been "mouthpiece to his majesty, Joseph Henry."

Temporary detachment from the society of the learned seemed to have awakened his interest in common humanity. He turned his attention from the works of nature to the ways of man. "There are so many curious customs, such as eating on the street, women sitting on the sidewalks in groups, sewing and talking as freely as if they were in their own houses, men harnassed to carts, drawing stone or other articles about the streets, the general employment of women as clerks and shopkeepers, women engaged in many occupations followed only by men in America, such as watch-making, street-cleaning, and even shoe-making. One is struck both in London and Paris with the number engaged in unproductive labor, earning a scanty subsistence by exhibiting feats of strength, or by passing through the streets picking up rags, bits of paper, or other discarded articles."

He expressed his distaste for the prevailing military spirit. It was not so much the martial display which took place before the hotel each morning for the edification of a colonel who occupied an apartment on a lower floor. What seemed particularly to annoy him was the obligation of the National Guards (and this embraced every male between the ages of eighteen and sixty) to serve one day each month on military duty. College professors were not exempted.

The arrival of American newspapers acquainted him of the extent of the financial depression into which the nation had sunk. He commented on the evil causes and consequences in a letter to Bache, who was still in England. "It is absurd to attribute it to the idea of the administration. The distress commenced in England, and was produced in our country by the ruinous system of paper money and the consequent spirit of gambling and speculation which, for two years, has infected everyone. Instead of applying our minds in the production of objects of real value, we have been entirely engaged in making paper fortunes, or rather in inflating a bubble which at length has burst. But you are beginning to laugh; it is no matter for amusement."

Henry was having financial troubles of his own. He had already bought apparatus for the college laboratory to the amount of 800 dollars and, in a letter to Stephen Alexander, he does not conceal his dismay at the news he was to be allowed only 500 dollars for this purpose, although he had been promised more.

Once Henry had polished up his knowledge of French with the aid of a native instructor, and had acquired the services of Thomas Hun as interpreter, he began to seek the society of the scientists whom he knew by reputation. He was disappointed at the progress he made. There was no lack of cordiality in their welcome, but everyone was so busy that he could not enter into friendships similar to those he had formed with the scientists in London. "The French savans of the older class are so much occupied in the business of teaching, also in politics and legislation (they are all members of the Chamber of Deputies) they have no leisure for attention to strangers. I have, however, got into a different current, having picked up much valuable knowledge among a younger class. They have been very kind to me and will form valuable acquaintances in the way of correspondents."

His most conspicuous acquaintance was Macedoine Melloni, famous for his studies of the laws of radiant heat and who, at the age of thirty-five, was already a member of the Institut des Sciences. He also met De La Rive (passing through the city on his way from Geneva to London), Gay-Lussac, Biot, and others. He was gratified by what he learned from them of their methods of teaching and instruction.

He attended a meeting of the Institut but was astonished to observe that the members gave more attention to their newspapers than they did to the paper which was being read.

The Paris visit was not without its adventure. Henry had gone to view the festivities in connection with the marriage of the Duke of Orleans, the king's son, during which a stampede occurred in which several people were trampled to death. Henry was exposed to considerable danger of injury or worse, since he was within a few feet of the entrance toward which the mob surged in its panic. Only by exerting all his strength was he able to extricate himself from the danger of being thrown down and trampled on.

His original plans had embraced a visit to Geneva and a short tour of the Low Countries, but the hospitality he had met in London had exceeded his expectations and had induced him to prolong his stay in that city. Paris, too, proved more attractive than he had anticipated and he overstayed his scheduled time there, so that he was obliged to revise his program. There was still an unaccomplished portion of his English visit which he felt could not be foregone. Accordingly, he left Paris on July 12, and after more than twenty-eight hours of almost continuous driving he arrived in Brussels. A long letter home described the pleasant aspect of the countryside, the Belgian capital, and the field of Waterloo.

After three days of sightseeing he went on to Antwerp, where he was enchanted with the resemblance to his native Albany. The brick houses with their tall gables made him feel at home and satisfied him that the Dutch settlers had carried the imprint of their native land into the New World. He abandoned his trip to Holland. On July 24 he was back in London, where he took up residence with his former student Henry James, at 68 Albany Street.

He did not immediately seek to renew the acquaintances he had made, but devoted himself to the welfare of his friend James, who had fallen into a state of lethargy. For three months the young man had closed his door to all friends and had buried himself in books. Henry succeeded in arousing young James to leave his books and to resume a normal intercourse with his friends. When Henry began to renew his own acquaintances with scientific society he found his association with James a source of minor embarrassment.

James was accompanied by a colored servant who made himself agreeable among the other servants with whom he was thrown into contact. This man became the cause of much comment at the dinner tables to which Henry was invited, and the visitor was put on his mettle defending slavery in America. In defense, he pointed out the inequality of privileges enjoyed by the different classes in England, but he recognized the inadequacy of the parallel and did not attempt to justify the slave system.

Resumption of his relations with fellow-scientists was interrupted by an indisposition which kept him in bed for a few days, but on his recovery he called again upon Faraday and was warmly received. To celebrate the renewal of their acquaintance, Faraday took his visitor on an excursion on the newly opened London and Buckingham Railroad, which ran a distance of twenty-five miles in the general direction of Birmingham. Henry was delighted when he was permitted to ride on the locomotive, and was much impressed by the engineering feat performed in excavating the two long tunnels through which the train passed.

On August 3, Henry resumed his travels, leaving by steamboat for Edinburgh. Like many another traveller before and since, he found the weather in the Scottish capital both chilly and wet, but he found compensation in the warm greetings awaiting him. He soon made himself known to the leading scientists of the famous old university, but in his long letters home he devoted most space to a description of his visits to lighthouses. Curiously enough he was to be closely associated with American lighthouses before many years had passed.

If he had been frankly disappointed by the architecture of London, his delight in architecture had been somewhat restored while he was in Paris, but he now thoroughly enjoyed the general aspect of Edinburgh, whose inhabitants proudly call it "the Athens of the North." Since both Harriet and he were of Scottish descent and the Scot's attachment to the parent land is proverbial, he enlarged on the beauty of the city and its historic places to great length in his letters home. His comments reveal the wide knowledge he possessed of Scottish history and literature.

After two weeks of pleasant visiting, he called upon Sir David Brewster, who had generously taken notice of his work in the *Philosophical Journal* which he edited. Brewster had also spoken in admiration of Henry's work in his latest article on magnetism written for inclusion in the new edition of the *Encyclopaedia Britannica*. Henry had a high opinion of his host's scientific attainments, especially in the study of light. He was delighted by the friendliness displayed by the Scottish scientist. He was flattered, too, when Sir David demonstrated some unpublished experiments with the dark lines in the solar spectrum.

Lady Brewster was a daughter of the McPherson who had written the poems of Ossian, with which Henry was familiar, and it was only natural that he should have revelled in the literary atmosphere of this hospitable house. He was further able to meet the original Dominie Sampson of Scott's *Heart of Midlothian,* and did not fail to make the pilgrimage to nearby Abbotsford, the home of Sir Walter. But the literary tid-bit of the visit was a letter in Brewster's possession written by Sir Isaac Newton to a widow of his acquaintance. Any letter in the handwriting of "the prince of philosophers" would have stirred Henry's pulse (he had thrilled with excitement when he had handled the original manuscript of Newton's *Principia*), but this was unique and unconventional. The letter set forth in a series of semi-mathematical propositions how it was the lady's duty to forget the memory of her departed husband in the arms of the living scientist.

Loath to leave such interesting hosts, Henry delayed his departure until he barely had time left to hurry through a round of calls upon his mother's relatives in Ayrshire. From here he went on to Liverpool, where he hoped to have a final reunion with the many friends he had made in Europe who were assembling for the annual meeting of the British Association for the Advancement of Science. He was invited to address the physics section and responded somewhat briefly with an account of experiments he had performed on the lateral discharge of Leyden jars,[21] a phenomenon he had been led to investigate through some comment of Roget's. He did not add material to the subject but he recommended it as a fruitful field for further research.

Henry's presence at this meeting gave rise to a remark that has become historic. He had appeared at a meeting of the section on the Mechanics of Engineering. He was recognized by the chairman, who invited him to inform the members on the progress made with the steam engine in the United States. In the course of his talk, Henry spoke of the vast improvements made in water transportation by the development of the steamboat which, he said, had been made to travel at a speed of fifteen knots.

In the discussion which followed, a Dr. Lardner, who enjoyed a reputation as an economist, made some comments of a derogatory

---

[21] Brit. Assoc. *Report for 1837.* Vol. VI, pt. 2, p. 22; *S.W.* Vol. I, p. 101.

nature in which he saw fit to brand Henry as a foreigner and as a representative of that nation which was notoriously prone to exaggeration. As a result Henry concluded it was hazardous for an American to make any communication to a learned society unless he was of established reputation. He acknowledged that the affair was unpleasant at the time, but that it turned out happily in the long run because Lardner's ill-mannered remarks were so resented by those who learned of the incident that everyone doubled their efforts to do the victim of the attack a kindness.

Henry had ample revenge on his opponent.

Dr. Lardner had flatly declined to believe that steamboats could be made to travel through the water at a speed of fifteen knots, as Henry had claimed. He then proceeded to make his encounter with Henry immortal by adding: "As to the project which is announced in the newspapers of making the voyage directly from New York to Liverpool, it was, he had no hesitation in saying, perfectly chimerical, and they might as well talk of making a voyage from New York or Liverpool to the moon."

Six months later two vessels steamed into New York harbor, having completed the voyage from England entirely under steam!

Lardner's false prophecy has found its way into the histories of trans-Atlantic travel, but it is well to remember it was Henry's pride in American accomplishment which provoked it.

The European journey had now come to its end. He was anxious to return to his family. The first available packet to sail was the *Toronto,* which made the passage in twenty-six days. It was not a comfortable voyage, for Henry was wretchedly seasick for the first week. He recovered in time to learn from an old mariner some curious facts about "a faint light that could be observed at the ends of masts and spars and that this on examination was found to proceed from a gelatinous substance of the phosphorescent kind." Strange, that the man who knew so much about the new science of electricity did not have a better acquaintance with St. Elmo's fire, one of the oldest known forms of electricity.

Once more on American shores, he was promptly robbed of his journal containing the record of his Scottish journey.

On his return, Henry expressed the feeling that he was now better qualified to give an account of the state and the teaching of science in Europe than any other visitor in recent years. He had profited, too, by adding to his knowledge of the experimental methods of the better equipped Europeans. He came back in excellent condition, strong and fervent, charged with enthusiasm to resume his own investigations.

Shortly after his return from Europe Henry was to meet two of the men who, with very different fortunes, were making practical applications of the principles he had discovered. The first meeting was with Thomas Davenport, the Vermont blacksmith who had bought one of the Penfield Ironworks' magnets and who had vainly sought to meet Henry in Albany. Having experimented with the magnet he had acquired, Davenport had succeeded in developing a rotary electric motor of crude design and was seeking financial support to exploit his invention. He went to Princeton in order to obtain Henry's approval of the motor. Since Davenport's invention was in a way confirmation of Henry's assertion that his simple reciprocating motion machine provided a basis for further development, he welcomed the inventor, giving him freely of his time and advice. He advised Davenport to continue experiments for the improvement of the motor. He also recommended the inventor to build his apparatus on a smaller scale for two reasons. First, it would reduce the expense and second, because a large model looked too commercial. The second reason showed that Henry was not entirely without a shrewd touch for, he said, if the story got about that a large motor developed defects that would be the end of it, whereas a small motor of an experimental nature could fail and its failure could be passed off with a shrug of the shoulders and a promise to benefit from the experience.

Henry gave Davenport the recommendation he wanted and provided him with letters of introduction to Bache and Eaton, both of whom had influential friends who might be interested in the commercial possibilities of the motor. Unfortunately, Davenport was too far in advance of his times, he failed to persuade anyone to finance his undertaking, and he sold the motor eventually to Amos Eaton who wheedled the money out of General van Rensselaer.

The second inventor whom Henry met at this period was to encounter much better fortune than the unappreciated Davenport. This was Samuel Morse, who claimed to have conceived the idea of the electric telegraph in 1832 while returning from England on the *Sully*. Morse was undergoing the cruel experience of learning that it is one thing to conceive an idea but quite another thing to make it practical. An artist of no mean ability but entirely without scientific training, unaware that others had conceived the same idea and were working upon it with better qualifications to obtain success, Morse had set about the realization of his idea. When the telegraph he built failed to operate he realized his technical shortcomings and consulted a colleague, Leonard Gale, professor of chemistry at the College of New York. Gale instantly saw Morse's errors and recommended a thorough study of Henry's paper on the transmission of electrical impulses over a long wire which had appeared in Silliman's *Journal*. This paper and further assistance from Gale set Morse on the right track.

After his return from Europe Henry was informed that Morse was about to petition Congress for financial assistance to construct his electric telegraph system. Friends urged Henry to take the initiative by informing Congress that the publication of the physical principles had put the electric telegraph in public domain. They urged him to recommend Congress not to make an appropriation to any individual to promote a system but, instead, to offer a substantial reward to the man who produced the best practical system. Henry intended to follow this advice but in accordance with the habit which was the despair of his friends he delayed taking any action until he had given the matter further thought. Then he met Morse and his attitude was changed.

While on a visit to New York he called at the chemical store of a Mr. Chilton, on Broadway, to obtain supplies for the college. Chilton was well informed on Henry's work and had already discussed it with Morse. While Henry was in his store on this occasion Morse walked in and the chemist made them known to each other. Morse impressed Henry as "an unassuming and prepossessing gentleman with very little knowledge of the principles of electricity or magnetism. He made no claims in conversation with me to any scientific discovery beyond this particular machine" he was working on and which failed to operate.[22] Their conversation in Chilton's store served to delay writing the letter to Congress on the question of an appropriation to encourage the development of the telegraph. Henry was of opinion that if he acted under Gale's guidance Morse deserved to succeed. In the year 1839 Morse had returned from Europe discour-

[22] *Smithsonian Report for 1857*, p. 114.

aged by his reception in that country where Wheatstone's needle telegraph was highly regarded. Finding that Gale was absent in the South but desiring some technical advice, he wrote to Henry for an appointment. To understand certain allusions in his letter it is necessary to know that Henry had recently sent him a copy of *Contributions to Electricity,* number iii, and that Henry had borrowed from Gale five miles of copper wire which belonged to Morse.

Morse wrote: ". . . In case I should be able to visit Princeton for a few days in a week or two, how should I find you engaged? I should come as a learner and could bring no 'contributions' to your stock of experiments of any value, nor any means of furthering your experiments except, perhaps, the loan of an additional five miles of wire which it may be desirable for you to have. I have many questions to ask."

In his reply Henry urged the inventor to persevere with his efforts to complete the electric telegraph. He said: "I believe that science is now ripe for the application, and that there are no difficulties in the way but such as ingenuity and enterprise may obviate. But what form of the apparatus or application of the power will prove best can, I believe, be only determined by careful experiment. I can only say however, that, so far as I am acquainted with the minutiae of your plan, I see no practical difficulty in the way of its application for comparatively short distances, but if the length of the wire between stations is great I think that some other modification will be found necessary to develop a sufficient power at the farther end of the line. I shall, however, be happy to converse freely with you on these points when we meet." [23]

A modification which Henry was willing to talk about freely proved to be the insertion of the relay in Morse's system. This would remove the limitation in distance over which messages could be telegraphed. Henry had already communicated his knowledge of the device to Wheatstone, who had incorporated it in his needle telegraph.

When they eventually met Henry found the time necessary to satisfy his visitor's needs and gave Morse all the information he possessed upon the telegraph.

A glimpse of Henry's warm nature was recounted in a recollection from this period of Dr. S. Weir Mitchell, the novelist and physician.

[23] Morse, E. L. *Samuel F. B. Morse: his letters and journals.* Vol. II, p. 140.

When a boy of fifteen, Mitchell was sent by his father to Professor Henry with some glass apparatus which could not be sent otherwise for fear of breakage. The professor met the lad at the railroad station, took him home, and entertained him overnight. Next morning Henry spent much time in explaining his experiments in transmitting signals over wires. His young guest was overwhelmed by the kindness bestowed upon one so young by a famous man and expressed himself effusively at the time of his departure. His host could not refrain from smiling at the profuse flow of thanks but in setting the boy at ease he put into words a fragment of that philosophy upon which his relations with other men were based.

"Well, life sometimes gives one a chance to return little favors," he said, "and perhaps some day you will have an opportunity to oblige me."

The day came, long after, when Mitchell was to remember his indebtedness.

After his return from Europe, Henry resumed his researches into electromagnetism. In the investigations which now followed there is imagination of a very high order. For ingenuity, completeness, and novelty in the examination of new phenomena these researches serve as a model for the experimental physicist. Their significance lies, not so much in the material which he added to the science, as in the vast territories he opened to the view of those who followed in his steps.

The four stages of scientific research are observation, interpretation, prediction, and realization. The whole process must be controlled and illumined by prepared imagination. The qualifying term "prepared" is essential, because the imagination of the poet who calls the stars the windows of heaven is beautiful, but it is not useful in the sense that imagination is employed in scientific research.

It is difficult to conceive of any great scientist who is not also an artist or a poet. In his youth Henry had displayed this side of his nature in his love of novels and plays. He had his romantic moods. At that time of his life imagination was dominant. Now it was to be subordinated to other qualities, but it was employed along a direct path that led to the revelation of great truths.

Henry resumed his study of the induced current from a point of view which was revealed in one of his later papers. "The secondary currents, as it is well known were discovered in the introduction of magnetism and electricity, by Dr. Faraday, in 1831. But he was at that time urged to the exploration of new, and apparently richer veins of science, and left this branch to be traced by others. Since then, however, attention has been almost exclusively directed to one part of the subject, namely the induction from magnetism, and the perfection of the magneto-electrical machine; and I know of no attempts, except my own, to review and extend the purely electrical part of Dr. Faraday's admirable discovery." [24]

From this quotation, as Magie points out, "it may be conjectured that Henry did not appreciate that the production of the secondary current by the movement of magnets and by setting up or breaking a primary current were simply variants of the same action." [25]

Henry undertook a series of elaborate experiments which extended over several years. The results were presented to the members of the American Philosophical Society under the general heading of "Contributions to Electricity and Magnetism."

There was now no longer need for any admonition to induce Henry to announce the results of his experiments. The congenial company in which he found himself at the meeting of the American Philosophical Society evoked a steady flow of papers, verbal communications, and letters, which were subsequently recorded in the Society's *Proceedings*.

On February 16, 1838, he amplified his observations on the lateral discharge of Leyden jars with the comment that, even with a good "earth," free electricity is not conducted silently to the ground. Three months later (May 4) he sent a letter to tell the members that, on the discharges from a Leyden jar through a good conductor, a secondary shock could be obtained through a perfectly insulated nearby conductor that was more intense than the shock delivered directly from the jar itself.

These were only preliminary studies conducted in the intervals between college duties, but they were leading to an intensive study of the secondary current that had received so little attention during the seven years since its discovery had been announced. During the summer vacation it is evident the professor found the leisure to make

[24] *S.W.* Vol. I, p. 114.
[25] Magie, W. F. *Op. cit.*, p. 482.

more extended investigations, for at the meeting of the Philosophical Society on November 2, 1838, he delivered a paper which gives a number of experiments in detail.

Judged by any standard this is a remarkable paper. It is generally conceded to deserve distinction because it describes the discovery of induced currents of the third, fourth, and fifth orders. But not the least noteworthy feature is his clarification of the principles of the transformer. Both Faraday and Henry used a transformer construction in their first mutual induction experiments, but it was not until the publication of this paper that Henry demonstrates how, by the proper proportioning of the coils, the voltage could be stepped up or down with this structure. This is fundamentally the basis of all modern electrical transformers.

It is quite apparent that he planned these experiments with a great deal of foresight, for he prepared a number of flat coils of copper ribbon about 1½ inches wide, and insulated with two coatings of silk. Most of the coils contained 60 feet of the copper ribbon, but one was 93 feet long. Various coils of wire, called helices in the paper to distinguish them from the ribbon coils, were also used. These were wound so that one would fit inside the other to permit extension in length when desired. This arrangement permitted him to form a helix of 3,000 yards. The spool of copper wire, 5 miles long, belonging to Morse, also figures in the experiments.

To determine the direction of the current he used a magnetizing spiral consisting of a small number of turns of copper wire wound around a straw or thin glass tube in which a sewing needle was inserted.

He also used a small horseshoe bar wound with a few turns of wire, so as to constitute a quantity magnet.

In the earlier experiments he used an ordinary zinc-copper battery in a single cell, in which the active surface was about 1¾ feet of zinc. Later, he substituted for this a set of Daniell's cells, which gave him a steadier primary current.

The first set of experiments to be presented relate to the properties of the extra current. By using the longer helices he was able to find that the current produced when the circuit was broken was of great intensity, even when the battery power was small. He gives one striking example.

A very small compound battery was formed of six pieces of copper bell-wire, about one inch and a half long, and an equal number of pieces of zinc of the same size. When the current from this was passed through the five miles of the wire of the spool, the induced shock was given at once to twenty-six persons joining hands.[26]

In the second section of the paper he considered the secondary current. He observed that when the inductive action took place between one of his copper coils used as the primary, and a similar coil used as the secondary, the needle in the magnetizing spiral was strongly magnetized. Magnetism was exhibited by the small magnet and, when the ends of the secondary were attached to a decomposing apparatus, gas was generated at each pole. The shock from this coil, however, was feeble. The induced currents produced in this case had considerable quantity but low intensity. By using a longer coil as the secondary the induction of quantity did not increase, and in some cases diminished, but the shock they gave was more pronounced.

A helix was then substituted for the ribbon coil as the secondary. The decomposition produced in the electrolytic cell by the current was much less than with the coil, and the sparks were smaller, "but the shock was almost too intense to be received with impunity, except through the fingers of one hand. A circuit of fifty-six of the students of the senior class, received it at once from a single rupture of the battery current, as if from the discharge of a Leyden jar weakly charged. The secondary current in this case was one of small quantity but of great intensity." [27]

Variants of this experiment were tried. When a wire helix was made the primary and one of the copper ribbon coils served as secondary, there was obtained an induced current which gave no shocks, but which magnetized the small horseshoe magnet. The current obtained, therefore, was one of quantity and it was produced by the inductive action of a long helix from an intensive battery.

Henry summarized the results of these experiments as follows: "This experiment was considered of so much importance, that it was varied and repeated many times, but always with the same result; it

[26] *S.W.* Vol. I, p. 113.
[27] *Ibid.,* p. 116.

therefore established the fact *that an "intensity" current can induce one of "quantity,"* and, by the preceding experiments, the converse has also been shown, that *a "quantity" current can induce one of "intensity."* [28]

Here, then, is the electrical transformer, although few people trace it back to this source. It is one of Henry's notable discoveries.

In the midst of making these discoveries he does not disdain to tell his learned audience how he enjoyed the demonstration of a sober truth when it could be enlivened by a little mischief. Some unsuspecting members of the senior class were probably also the victims of the device mentioned in this passage:

> I may perhaps be excused for mentioning in this communication that the induction at a distance affords the means of exhibiting some of the most astonishing experiments, in the line of *physique amusante,* to be found perhaps in the whole course of science. I will mention one which is somewhat connected with the experiments to be described in the next section, and which exhibits the action in a striking manner. This consists in causing the induction to take place through the partition wall of two rooms. For this purpose coil No. 1 [29] is suspended against the wall in one room, while a person in the adjoining one receives the shock, by grasping the handles of the helix, and approaching it to the spot opposite to which the coil is suspended. The effect is as if by magic, without a visible cause. It is best projected through a door, or thin wooden partition.[30]

The summer vacation during which these experiments were made may have been charged with many surprises for the unwary visitor to Henry's laboratory, but one was found who was willing to explore this magic in search of its therapeutic value. In the next paragraph of his paper, he tells how a medical acquaintance, having a patient suffering from partial facial paralysis, induced the professor to give him galvanic treatment. The helix was suspended by a string passing over a pulley, and gradually lowered to the plane of the coil, until the required intensity of the shock was reached. Unfortunately, no

---

[28] *Ibid.*, p. 118.
[29] Thirteen pounds of copper ribbon, 93 feet long, and 1½ inches wide.
[30] *S.W.* Vol. I, p. 121.

record of the progress of the patient was appended to the report. The poor man was no more than a part of Henry's experimental apparatus which, in this case, may be recognized as a very crude form of variometer, an instrument now widely used.

The fourth section of this paper is concerned with the screening effect found when a sheet of conducting material was interposed between the primary and secondary coils. Both Henry and Faraday investigated this shielding effect. That its meaning was grasped by Henry seems to have been due to his method of observation which indicated peak voltages, in contrast to Faraday's, which gave only a measure of the total quantity of induced electric displacement.

Before he began his experiments, Henry mentioned that he knew of the discovery of magnetic rotation by Arago, and of the experiments performed by Davy, Babbage, Herschel, and Harris. These experiments showed that a revolving magnet gives rotation to metallic discs or needles suspended above it, and that this action is stopped by an interposed screen of magnetic material or thick plates of copper, silver, or zinc. Faraday had also shown that the rotation is due to inductive rather than to magnetic action, the rotating magnet caused induced currents in the suspended body.

Henry was not aware that Faraday had made these experiments, since they were not available to him until the *Philosophical Transactions* for 1838 were received in this country. In the course of his work Faraday had found that a copper plate, when interposed between the primary and secondary coils, effectively shielded the secondary from the magnetic effects of the primary. Under this condition, the induced currents in the shield, which set up a current opposing that of the primary coil, die out before the galvanometer response is well under way, so that its deflection is dependent only on the total change of current in the secondary coil, and hence are substantially independent of the screen.

In Henry's investigation of the screening effect of materials placed between two coils, he found that conductors destroyed the induction, while non-conductors did not. He proved that the screening effect of the conductors was due to the induction in them of eddy currents (although he did not use that term) which neutralize the current induced in the screened coil. He obtained this proof by cutting a triangular slit in the metal screening disc. The slit disc did not exert any screening effect. Henry then traced the direction which the current would have taken in the complete disc by connecting the edges of the slit by wires to a magnetizing spiral with a needle in it.

He summarized his investigation of the phenomenon as follows: "All good conducting substances are found to screen the inducing action, and this screening effect is shown, by the detail of a variety of experiments, to be the result of the neutralizing of a current induced in the interposed body." [31]

He then went on to point out that this principle had an important bearing on the improvement of electrical machinery, since the eddy currents induced in homogenous conductors must reduce the efficiency of these machines, and stated his belief that advantages would be found in using bundles of wire instead of solid pieces of iron. Today, laminations are used.

During the meeting of the Philosophical Society at which this paper was presented Alexander Bache gave an account of an investigation conducted by Professor Ettinghausen of Vienna, in which he mentioned that he had been led to suspect the development of a current in the metal of the "keeper" of his magnetoelectric machine, which diminished the effect of the current in the coil about the keeper. Henry quickly grasped the significance of this announcement and said: "It gives me pleasure to learn that the improvement which I have already suggested as deductions from the principles of the interference of induced currents, should be in accordance with the experimental conclusions of the above named philosopher." [32]

When Henry later learned of Faraday's experiments and of the totally different results obtained, he was embarrassed. However, he instantly opened a new series of experiments and, from the results he obtained and a clear piece of reasoning, he was able to arrive at a reconciliation of their divergent views. These experiments were described in a later paper.[33]

Both men were aware of the fact that there was one quality of the current which depended upon the rate at which the magnetic field is changed, this being what determined the shock in the muscles and the distance at which a spark will occur in a broken circuit, and that there was another quality of the secondary current depending

[31] S.W. Vol. I, p. 106.
[32] S.W. Vol. I, p. 145.
[33] S.W. Vol. I, p. 169.

upon the total change of primary current, which determined the throw of the galvanometer needle. The former was independent of the material of the conductor, while the latter varied with it. Henry was first to grasp this difference.

Reading the description of the two men's investigations, one arrives at the impression Henry had the clearer conception of the essential features of induced currents, but no conclusive proof for such a belief can be advanced. The impression may be the outcome of his clearer and more forceful manner of presenting his views. The work of the two men on screening effects was not advanced until thirty years later by Clerk Maxwell.

However, one perceives Henry's tenacity of purpose in these experiments. He had made an observation that had been disputed. Nothing would satisfy him until he had gone beyond the mere proof that his own assertion was correct, he had to have a thorough understanding of the discrepancy in one or other of the observations.

The observation of the creation of eddy currents which neutralized the primary seems to have suggested to him a new course for investigation. If a secondary current could be produced from a primary, could it be used to induce a third current in a conductor?

During the seven years that electrical induction had been known, no one had thought to determine this point. It was by no means certain that a tertiary current would be produced, for the current in the secondary being only momentary it might be suspected that the effects of its growth and decay might come so close together in time as to annul each other. This turned out not to be the case.

He joined two coils together in the same plane, putting one of these coils as secondary over the original primary, and putting the third circuit over the second coil of the secondary. When the test was made, a current was developed in the tertiary which gave an intense shock. He related how, with only a small battery power, he was able to pass this shock through twenty-five students forming a circuit. Once more we have the picture of the professor's glee at the sight of his students writhing under his galvanic administrations when they consented to serve as his measuring instrument.

By extending the arrangement of coils, currents of the fourth and fifth order were obtained. Consideration of these currents of higher orders and of their relation to the original current led him to the belief that they must consist of a succession of currents alternating in direction. On making tests to determine in which direction the currents flowed, he learned that this was so.

Henry was led to the next step in these experiments by the consideration: "the fact that the secondary current, which exists but for a moment, could induce another current of considerable energy, gave some indication that similar effects might be produced by a discharge of ordinary electricity, provided a sufficiently perfect insulation could be obtained." [34]

To test this notion he pasted a narrow strip of tin foil spirally around the outside of a hollow glass cylinder, and another similar strip around the inside of the cylinder, so that the spires of the two were directly opposite to each other. When the ends of the inner ribbon were joined through the magnetizing spiral and a discharge from a Leyden jar was sent through the outer ribbon, the needle in the magnetizing spiral was strongly magnetized. When the ends of the inner ribbon were brought close together, a spark passed through the gap when the discharge was sent through the outer ribbon.

By using more cylinders in combination, currents of the third and fourth order were obtained.

After describing in detail a number of experiments, Henry concluded that "from these experiments it appears evident that the discharge from a Leyden jar possesses the property of inducing a secondary current precisely the same as the galvanic apparatus."

This splendid paper, embodying so many original ideas, gave promise for a succession of other papers which would continue to advance the knowledge of electricity. But when he presented his fourth paper in the series of "Contributions" he had to express regret that his experimental work had been curtailed because of a temporary eviction from his laboratory while alterations were being made.

The fourth paper was read to the American Philosophical Society on June 19, 1840. It described the induction of the primary circuit. The experiments were performed with a Daniell's battery, and certain effects which had been considered important in earlier investigations were found to have resulted from the inadequacy of the battery power employed.

The manifestations of natural electricity known to the ancients were lightning, St. Elmo's fire, the attractive quality some substances dis-

[34] S.W. Vol. I, p. 133.

played when rubbed, and the discharge from an electric eel. That these were manifestations of the same force was not, of course, suspected until some understanding of the properties of electricity had been formed. Henry allowed his curiosity to be drawn to the least obvious feature of ancient knowledge, the electric eel.

How could an intense electric shock be produced in organs imperfectly insulated and immersed in a conducting medium?

Suspecting that the shocks might be the effect of a secondary current, he constructed an artificial eel. This was done by arranging a secondary wire coil furnished with two terminal handles, and a primary ribbon coil, the two coils being insulated. "By immersing the apparatus in a shallow vessel of water, the handles being placed at the two extremities of the diameter of the helix, and the hands plunged into the water, parallel to a line joining the two poles, a shock is felt through the arms." [35]

The riddle of the electric eel was explained.

As an addendum to this paper, Henry offered a theoretical explanation of the currents of higher orders.

The fifth and concluding paper in the "Contributions" was presented June 17, 1842, and although it contained one of Henry's most significant discoveries, the Philosophical Society did not print it in full. All we have is an abbreviated report in the *Proceedings*.[36] From this it appears that Henry presented a study of the peculiar anomalies observed in the direction of the currents of higher orders, and that he succeeded in tracing them to the oscillatory nature of the currents.

Although we are better equipped for research today than were the natural philosophers of a century ago, we are still unable to predict the polarity that will result in a bar of soft iron from a given condenser discharge. The French scientist Savary had magnetized needles by placing them within an activated helix of wire. He had learned that a frictional electric discharge did not always magnetize the needles with the same polarity. He concluded that the polarity was determined by the distance of the needle from the wire conducting the discharge. Henry had already observed this anomalous behavior, which some Germans accepted and referred to in their writings as "anomalous magnetism."

Henry could not accept the conception of anomalous magnetism.

---

[35] *S.W.* Vol. I, p. 161.
[36] Am. Phil. Soc. *Proceedings*. Vol. II, p. 193; *SW*. Vol. I, p. 200.

It was illogical. He therefore determined to make a series of investigations to arrive at the truth.

He made numerous attempts to observe the peculiar results by discharging a Leyden jar through a spiral. In tests with about one thousand needles of large size he found none with a reverse polarity. At last, using finer needles, he succeeded in obtaining changes in polarity simply by increasing the quantity of electricity while the direction of the charge remained unchanged.

Felix Savary had asked, "Is the electric movement during the discharge a series of oscillations transmitted from the wire to the surrounding medium and soon attenuated by resistances which increase rapidly with the absolute velocity of the moving particles?"[37]

Having greatly extended Savary's investigation, Henry confirmed his hypothesis by a clear and authoritative exposition of the phenomenon which reads in summary:

> The discharge, whatever may be its nature, is not correctly represented (employing for simplicity the theory of Franklin) by a single transfer of an imponderable fluid from one side of the jar to the other; the phenomena require us to admit *the existence of a principal discharge in one direction and then several reflex actions backward and forward, each more feeble than the preceding, until the equilibrium is obtained*. All the facts are shown to be in accordance with this hypothesis, and a ready explanation is afforded by it of a number of phenomena which are to be found in the older works on electricity, but which have until this time remained unexplained.[38]

In the light of this explanation the apparent changes in direction of magnetization of the needles is determined by the last oscillation having sufficient power to reverse the polarity of the needle. Were it otherwise, were the discharge unidirectional, the needle would always be magnetized to a degree of intensity proportional to the energy released; and it would be possible in every case to foretell with certainty the resulting polarity which the needle would acquire.

Unlike most of his discoveries, this was not overlooked by European scientists, although few associated it with Henry's name. Helmholtz,

[37] *Annales de Chimie et de Physique*. Vol. 34, 1827, p. 54.
[38] *S.W.* Vol. I, p. 201.

in his famous essay *Erhaltung der Kraft,* "assumes" the discharge of the jar to be oscillatory. Henry's hypothesis was confirmed by the experiments of Fedderson, but it remained for Sir William Thomson (Lord Kelvin) to furnish the mathematical proof and to give Henry's discovery the prominence it deserved.[39]

This was not the only discovery Henry had to report. The magnetizing spiral he was using to polarize the needles proved to be a sensitive recorder of induced current. For example, he says:

> A single spark from the prime conductor of the machine, of about an inch long, thrown on the end of a circuit of wire in an upper room, produced an induction sufficiently powerful to magnetize needles in a parallel circuit of wire placed in the cellar beneath, at a perpendicular distance of thirty feet with two floors and ceilings . . . intervening. The author is disposed to adopt the hypothesis of an electrical *plenum,* and from the foregoing experiment it would appear that the transfer of a single spark is sufficient to disturb perceptibly the electricity of space throughout at least a cube of 400,000 feet of capacity; and when it is considered that the magnetism of the needle is the result of the difference of two actions, it may be further inferred that the diffusion of motion in this case is almost comparable with that of a flint and steel in the case of light.[40]

Comparable it is indeed.

Henry was here delineating the approach of the electromagnetic theory of light which Faraday advanced four years later in his essay on "Ray Vibrations." Henry suspected a similarity between electrical effects and light, which was subsequently proved by Clerk Maxwell's brilliant work in 1863, when he demonstrated mathematically that light waves behave like electromagnetic waves, and that they are both transmitted through space in the same manner and with the same velocity.

One can understand that Henry might have experienced some difficulty when he made the observation that electromagnetic and light waves were comparable. Whatever suspicions he may have nourished upon their analogy would run counter to the popular belief of

---

[39] Transient Electric Currents. *Philosophical Magazine.* June, 1853.
[40] *S.W.* Vol. I, p. 203.

Newton's corpuscular theory, upon which he had been trained. This theory had not yet yielded to the wave theory of light based on the experiments of Young and Fresnel. Maxwell's theory was a beautiful vindication of their, as well as of Henry's, suggestion.[41]

But before Maxwell published the mathematical proofs, Henry had arrived at a clear perception of electromagnetic waves and their behavior. In the year 1851 he presented a paper before the American Association for the Advancement of Science on the "Theory of the So-called Imponderables," in which he wrote that the inductive effects of the discharge currents from a Leyden jar "are propagated wave-fashion." In the same paragraph he says that these inductive effects extend "to a surprising distance; and as these are the results of currents in alternate directions, they must produce in surrounding space a series of plus and minus motions and analogous to if not identical with undulations." [42]

When Maxwell gave credit to Faraday for having proposed the electromagnetic theory of light in 1846, he probably was not aware of Henry's prior suggestion. It is strange, however, that Faraday did not give it the consideration and mention it merited.

That Henry elaborated upon the subject before his class is clear from the notes on "Lectures on Natural Philosophy by Professor Henry" made by one of his students, William J. Gibson. These are dated February 28, 1844. After a description of experiments on induction at a distance, the student wrote in his note-book:

> Hence the conclusion that every spark of electricity in motion exerts these inductive effects at distances indefinitely great (effects *apparent* at distances of one-half mile or more); and another ground for supposition that electricity pervades all space. Each spark set off from the Electrical Machine in the College Hall sensibly effects the surrounding electricity through the whole village. A fact no more improbable than that light from a candle (probably another kind of wave or vibration of the same me-

---

[41] Although it has been generally assumed that Newton was a confirmed believer in the corpuscular theory of light, Jeans has recently pointed out that this view is erroneous. (Jeans, Sir James. *Growth of Physical Science*. Cambridge, 1948. p. 207.) Newton explicitly stated that he preferred another theory, and this is a mixture of the corpuscular and undulatory theories. Henry either had a clearer view of Newton's thought, or he was sacrificing his Newtonian convictions when he expressed belief in the wave theory of light.

[42] A.A.A.S. *Proceedings*. Vol. VI, p. 84; *S.W.* Vol. I, p. 302.

dium) should produce a sensible effect on the eye at the same distance.[43]

Henry continued his experiments on induction at a distance after presenting this view. His own choice of an example best illustrating the nature of his work in this regard was revealed in a letter to his friend Rev. S. B. Dod. "It may be mentioned that when a discharge of a battery of several Leyden jars was sent through the wire before mentioned, stretched across the campus in front of Nassau Hall, an inductive effect was produced in a parallel wire, the ends of which terminated in the plates of metal in the ground in the back campus, at a distance of several hundred feet from the primary current, the building of Nassau Hall intervening. The effect produced consisted in the magnetization of steel needles." [44]

Incidentally, he was under the impression he had made the first use of earth returns in these experiments, for in the same memorandum he wrote: "I think the first actual line of telegraph using the earth as a conductor was made at the beginning of 1836." Is this another example of his failure to make known the introduction of an original idea? Subsequently he was compelled to acknowledge Steinheil as the discoverer of earth returns.

The effect of lightning on an electroscope had been observed many times before Henry's work on induction at a distance. In extending the investigation, Henry connected a wire to the metallic roof of his study, carried the wire through a magnetizing spiral to the ground, by means of which arrangement needles were magnetized in his study. "With every flash of lightning, which took place in the heavens within a circle of twenty miles around Princeton, needles were magnetized in the study by the induced current developed in the wire.[45]

Nor did he fail to note that "a series of currents, in alternative directions was produced in the wire."

It required the complete work of Clerk Maxwell and Heinrich Hertz to draw the full significance from this series of experiments. As our quotations have shown, he was not only making experiments with radio waves, but he was also beginning to formulate qualitatively

---

[43] Manuscript in Princeton University Library, p. 135.
[44] *Memorial,* p. 152. The distance was about 220 feet.
[45] *S.W.* Vol. I, p. 250.

some crude ideas of an electrical ether which transmitted disturbances to great distances.

Why, having done so much, did Henry not do the little more to develop the electromagnetic theory of light, or to lay a wider, firmer basis upon which a system of radio communication could be erected?

Henry's discoveries were isolated phenomena; there was nothing to which they could be related. His strength was always with experimental rather than with theoretical science, and he had advanced to the stage where experiment could hardly bear him farther. Before any advances could be made in that direction, some large gaps in knowledge had to be bridged.

Science progresses by the patient accumulation of facts, combined with bold guessing at the laws which bind the facts together. It required the leap of a genius of a different kind than Henry's to guess at what lay beyond the facts he had accumulated.

Henry had discovered that the discharge of a Leyden jar was oscillatory. More than a decade was to pass before Sir William Thomson succeeded in calculating the oscillations, and in making clear how the self-induction of a circuit, also discovered by Henry, may result in the production of alternating currents. He calculated in detail the nature of the oscillations—their period, intensity, and damping, as dependent upon the capacity, self-induction, and resistance.[46]

Still another decade was to pass before Maxwell's beautiful mathematical logic generalized and made more precise some of the other isolated phenomena. Maxwell built his electrodynamic theory upon sure foundations laid in experiments, first by Henry, and later by Faraday. His equations showed how electromagnetic waves proceed from electric oscillations. For fifteen years after the publication of Maxwell's complete conception, *Treatise on Electricity and Magnetism,* a great deal was written concerning the electromagnetic theory of light without advancing the subject. Physicists juggled with his equations but overlooked his ideas. Because the subject became a mathematical game instead of a scientific research the matter remained stationary.

Heinrich Hertz then opened the investigation which permitted him to prove the presence of these invisible waves and the actual existence of rapid oscillations in conductors excited by a spark discharge. He was able to prove the presence of nodes and loops of electric

[46] *Philos. Mag.* Vol. V, 1853, p. 393.

force in the wires by means of tiny electric sparks, and thus was able to determine the wave lengths. He further was able to produce stationary waves of this kind in free space between the oscillator and a reflector, and to measure their length. Yet, Hertz, who was scrupulously careful in acknowledging the work of other men in his field, nowhere mentions the name of Joseph Henry, except in regard to his work on sunspots.

Thus electromagnetic waves were found similar to light waves in every respect excepting length. While the waves of visible light, as measured by Fraunhofer and Fresnel, were of the order of a ten-thousandth of a millimeter, Hertz had found invisible waves measured in lengths of feet or in inches. No reason was left for doubt that the waves of visible light were similar to waves of electric force except in their wave lengths.

A new epoch in experimental and theoretical physics was opened. But it ought not to be forgotten that the impulse had been given fifty years earlier, when Joseph Henry had vaguely recognized what are now universally known as Hertzian waves, nor that he was the first to surmise there was a similarity between the emanation caused by the discharge of a conductor and the light emitted from an ordinary high-temperature source. Henry would have encountered difficulty in advancing to this stage, because he would have found Maxwell's theory difficult to comprehend because of the impossibility of devising a visual model which would relate its formulae to mechanical principles.

Recognition came tardily. Men who probably had never seen the journal in which Henry's results were published only became aware of the range, originality, and magnitude of his discoveries after his death, when his scientific papers were published and widely distributed by the Smithsonian Institution. In the year 1889, Sir Oliver Lodge wrote appreciatively in acknowledgement of Henry's discovery of the oscillatory nature of the Leyden jar discharge: "The writings of Henry have been recently collected and published by the Smithsonian Institution of Washington in accessible form, and accordingly they have been far too much ignored. The two volumes contain a wealth of beautiful experiments clearly recorded, and well repay perusal." [47]

Henry had missed fame through not publishing the results of his experiments in time; and when he reformed his dilatory habit and

[47] *Modern Views on Electricity*. London, 1889, p. 371.

became punctual in announcing his results, he remained inconspicuous because the vehicle he chose was inaccessible to European scientists. But once his scientific writings were made accessible, recognition of his achievements came at last. Sir J. J. Thomson paid tribute to his originality and to the value of his observation of the oscillatory discharge of the Leyden jar.[48] Still later, when Marconi was entertained at a dinner on January 13, 1902, in New York by the American Institute of Electrical Engineers, to celebrate the first transmission of a wireless signal across the Atlantic, he acknowledged his indebtedness to those who had paved the way for his accomplishment. "I have built very largely upon the works of others, and before concluding I would like to mention a few names—Clerk Maxwell, Lord Kelvin, Professor Henry, and Professor Hertz."

It was a noble company, and one in which Henry had earned his place.

In estimating the lasting effects of Henry's discoveries it is not enough that they should be numbered. They must be weighed before their value can be calculated. Only when this is done do we approach the measure of his worth, and of the fundamental nature of his contributions to the sum of human knowledge.

It was a tragedy that Henry's recognition as a great experimental physicist arrived too late to afford him any satisfaction. By the time the international scientific world awakened to a recognition of his merit he was already dead.

In America there was a growing interest in science, but there were too few scientists of repute. Although Henry was regarded as the foremost physicist in his own country, he could not understand why his work met with so little appreciation abroad. The low estimation in which American science was held may have been justified by the paucity of the original work being produced, but what neither Henry nor other American scientists could understand was the high regard in which Faraday was held while Henry met with small esteem.

Henry himself felt some resentment at this neglect. For all his modesty he was a sensitive soul, and he had a keen sense of what was due him. In order to correct in some measure the chilly silence upon his accomplishments, he wrote to Hans Christian Oersted at Copenhagen, drawing attention to himself much in the manner of an ad-

[48] *Notes on recent Researches in Electricity and Magnetism.* London, 1893, p. 332.

miring pupil inviting approval from a master who had overlooked his work.

Respected Sir: April 27th, 1841

My friend Mr. Steen Bille the chargé d'affaires from Denmark has kindly offered to transmit to you a small package containing some scientific pamphlets which you will please to accept, not on account of their value, but as a token of my respect for your scientific character. Science is of no country, and the discoveries you have made do not belong to Denmark but to the whole human family. Your name is as well known in the United States of America as in any part of Europe. You possess in every part of the civilized world the enviable reputation of being the founder of a new science, and all that has been done by others in the same line is but the exploration of the riches of the new region which you discovered.—I hope you will pardon the freedom of these remarks since they are the spontaneous expression of the feelings of one who has been, and still is, much devoted to the study of your science.

I regret that I am unable to send you a full set of my papers. I have copies only of the two last. My first experiments on electromagnetism were made in Albany, New York, in 1829–30. Of these you have probably seen an account. They were on the development of a great magnetic power in soft iron: and were made before Dr. Moll's experiments on the same subject were published. I was assisted in them by Dr. Ten Eyck.

I have invented a machine which moved by magnetic attraction and repulsion (see Silliman's Journal, vol. xx for 1831, page 340). This was the first movement of its kind ever made, but I have received no credit for it in Europe. Professor Jacobi in his publication [49] has never mentioned that I was before him in this invention. I next discovered, in 1832, the induction of a current on itself, or the means of getting shocks from a battery of a single element by means of a long conductor. I mention these results of my earlier labors because I think I have not received for them the credit in Europe which is my due.

You will find in the package a copy of the meteorological re-

---

[49] Jacobi, M. H. Ueber die Principien der electro-magnetischen Maschinen. *Poggend. Annal.* Vol. LI, 1840, p. 358.

port made by the Academy of the State of New York. These observations were established in 1825 by orders of the directors of the academies under the superintendence of Dr. Beck and myself; they have been continued ever since. If you are interested in them, I will have your name put on the list of persons they are presented, and you will then receive them annually through Mr. Bille. I have read with much interest and instruction your article on Electro-magnetism in the Edinburgh Encyclopaedia [50] and I have lately imported from Paris one of your articles of apparatus for exhibiting to my class in "Physique" your discoveries in the compression of liquids.

I have also read with much interest your paper on whirlwinds published in the Edinburgh Journal and also republished in America.

It would give me great pleasure to receive a letter from you. Any communication addressed to me through Mr. Bille will reach me safely.

With much respect, I am, yours, etc.

Joseph Henry.[51]

This is a curious letter, one which might be construed as the expression of a gnawing hunger for praise. However, there is no other evidence that such poison had entered Henry's mind, and much to the contrary. How many of us are willing to do good solid honest work, unpraised and unheeded?

Unfortunately, most of the letters addressed to Henry by his famous correspondents were accidentally destroyed, and it is impossible to tell what nature of reply this appeal for recognition evoked. There is reason to believe it may have been sympathetic, and that other letters followed, for Frederika Oersted mentions [52] that Henry at one time contemplated making a translation of the Danish scientist's book which appeared under the English title of *The Spirit of Nature,* but nothing seems to have come of this.

Most scientists are known for some single line of achievement. It is far otherwise with Henry. He never lost his capacity for new

---
[50] Published in America in 1831, and therefore too early to contain any allusion to Henry's work.

[51] Harding, M. C. *Correspondence de H. C. Oersted.* Copenhagen, 1920, Vol. 2, p. 383.

[52] *Breve fra og til Hans Christian Oersted.*

intellectual experiences; he always held open the door of his mind. In glancing over the subjects of his numerous scientific writings one perceives that his interests in science were almost universal. His greatest work was accomplished in the field of electricity, but he was by no means a specialist.

Not only does he appear to have been able to meditate upon almost anything with equal interest and intensity but, as he grew older, he displays an interesting duality of mind. Whenever he enters a new field of research he intuitively suggested the direction in which practical applications might be found, and when he studied a utilitarian project, as he frequently did, he enriched the basic science with illuminating observations. To an exceptional degree he was a combination of engineer and scientist. Although his qualities as an engineer were not fully revealed until the next stage of his career, we must make the comment parenthetically at this point so that the reader may observe this development in growth.

The great period of his labors in experimental science ended with the splendid work which brought him to the verge of a field in which the richest lodes had yet to be mined. One would have supposed that, with the vision of great possibilities opening before him, Henry might have desired to continue his researches into electromagnetism.

Whatever may have been his reasons for retiring from this type of research can only be surmised, but the fact remains he took a holiday from the subject in which he had earned distinction, and was never able to return to it. At a later date he summarized his work in this field in these words: "I have made several thousand original investigations on electricity, magnetism, and electro-magnetism, bearing on practical applications of electricity, brief mention of which fill several hundred folio pages. They have cost me years of labor and much expense."

None of his investigations in heat, light, or sound were productive of fundamental discoveries comparable with those he had contributed to electricity, but he did not fail to leave any subject the richer for his attention to it. Nor had he been gliding along in a well-worn groove as a teacher. He retained a strong belief in the practice of making scientific principles intelligible to those who had to use them. He did not fear to incur the reproach levelled against Humphrey Davy by a grave fellow-scientist, that his methods were "too lively." Henry would probably have accepted this as a compliment, for he never lost his zest for creating apparatus that could not fail to impress upon a student's mind an understanding of the principle it was designed to demonstrate.

One of his students, P. C. Van Wyck, graduate of the class of 1845, recalling some of Henry's demonstrations, said in a letter to the New York *Times:*

> While lecturing on the subject of "Sound" he would have a long pole about an inch in diameter passing from the basement to the philosophical hall in the attic, on which was screwed firmly a rude imitation of a fiddle at the top near the ceiling, while to the other end negro Sam had a real fiddle attached. On a bell signal from the Professor, Sam would saw away with his bow in the cellar, the Professor calling attention to the weird music his fiddle discoursed in the lecture-room. On these occasions Professor Henry always remarked that the function of the philosopher ceased when he demonstrated the principles of nature in his discoveries; that it then fell to the share of the inventor, by ingenious devices, to subordinate them to the use of man.[53]

After leaving electromagnetism, the first investigation Henry opened related to the phenomenon of phosphorescence, about which very little was known. The phenomenon was not new. Observation went back to the time of Pliny, but nothing had been added to the subject until the beginning of the seventeenth century when a cobbler of Bologna, Cascariolo by name, drew attention to the strange luminescence of barium sulphide. His observation served to promote a search for other substances which would become luminous under mechanical or chemical stimulation. John Canton studied the effect of the electric spark on sulphuretted lime, but the understanding of phosphorescent light made no real progress.

Then Henry learned that Becquerel, in France, about 1840–1841, had studied Canton's work in the light of more recent knowledge. The only outcome of Becquerel's investigation was the inference that the exciting cause of a luminous appearance of the lime was not identical with ordinary light.

Henry proceeded to repeat Becquerel's experiments before beginning any of his own. Then he searched for other phosphorescent

[53] Quoted in *Amer. Jour. of Science*. Vol. CXV, 1878, p. 465.

substances and confirmed the conclusions reached by Becquerel. Before he laid the study aside he had added forty phosphorescent substances to the list of those already known. He announced the results of his experiments at a meeting of the American Association for the Advancement of Science on May 26, 1843.

As a result of these experiments he expressed the view that "the exciting cause of the phosphorescence is an emanation possessing the mechanical properties of light and yet as different in other respects as to prove the want of identity." [54]

He found also that the phosphorescent rays excited by polarized light were completely depolarized. Although Henry announced this effect in 1843, it is usually attributed to G. G. Stokes who rediscovered it ten years later.

The most interesting of Henry's observations on phosphorescence was that the emanation differed from heat. The A.A.A.S. report said: "That the emanation also differs from heat is manifest from the fact that the lime becomes as luminous under a plate of alum as under a plate of rock salt, although these substances are almost entirely opposite in their property of transmitting heat. . . . In considering these emanations as distinct, he had reference only to the classification of the phenomena, for if they be viewed in accordance with the undulatory hypothesis they may all be considered as the result of waves, differing in length and amplitude, and possibly also slightly differing in the direction of vibration." [55]

From this it may be concluded that, if he still adhered to Newton's corpuscular theory of light, he was not unappreciative of the wave theory advanced by Huyghens, Fraunhofer, Young, and Fresnel.

Phosphorescence offers so many opportunities for the type of demonstration he delighted in performing that Henry could not resist introducing it to his students. One of his demonstrations was reserved for the occasions when the aurora borealis was active. He would then write on a sheet of paper with a solution of bisulphite of quinine. The written characters could not be seen under an ordinary lamp, but when the paper was exposed to the radiation from the northern sky, the characters glowed with a ghostly blue light.

On May 30, 1843, he read a paper to the members of the American Philosophical Society in which he broke new ground in the art and

[54] *S.W.* Vol. I, p. 208.
[55] *Ibid.*, pp. 208–209.

science of ballistics. In this paper, entitled "On a New Method of Determining the Velocity of Projectiles," he inaugurated the modern practice of employing electricity to time the flight of a projectile. The first apparatus he described consisted of a chronograph drum, revolving at a uniform speed of at least ten turns a second, and capable of regulation, so as to permit high-speed recording. Two screens of copper wire were placed in the path of the projectile whose flight was to be measured. The projectile in its flight cut through the screens, thereby interrupting an electric current passing through each; these circuit interruptions were automatically recorded on the surface of the drum by the deflection of the needle of a galvanometer.

Henry did not abandon the subject after reading his paper. Before his idea had time to appear in print he wrote to the Society's Reporter informing him that other methods had suggested themselves, but one in particular recommended itself to Henry and he requested that its description be included with the report of his paper when it was printed in the *Proceedings*.[56] This proved to be a decided improvement upon the original method, since it dispensed with the galvanometers. These were replaced by an arrangement which caused an electric spark to mark the paper wrapped around the recording drum. The recording paper, he recommended, should have a horizontal rotation that was also ascending. Then, by increasing the number of wire screens, the velocity of the projectile could be measured at different points in its path in a single experiment.

This method had much to recommend it, as it had no inertia in its working parts. It constitutes an invention in electrical engineering and in ballistics, and is the origin of the method which is now universally applied in measuring the velocity of projectiles in flight.

Another of his investigations made about the same time was inspired by the casual observation of mercury passing through a lead pipe. It had all the appearance of capillary action involving two metals —an unheard of thing.

This was much too strange a phenomenon to pass by without close investigation. Among the experiments he performed to verify his observation was one in which he bent a rod of lead, about one quarter inch in diameter and seven inches long, into the shape of a siphon. The shorter leg was immersed in a cup of mercury. After the lapse of twenty-four hours, a globule of mercury was observed at the end

[56] Amer. Phil. Soc. *Proc.* Vol. III, p. 165; *S.W.* Vol. I, p. 212.

of the longer arm. In the course of five or six days all the mercury had passed through the lead bar.

Pursuing these experiments, Henry was led to consult with men who were accustomed to work with metals, in order to ascertain whether other metals displayed similar characteristics. An experiment conducted with the aid of Dr. R. M. Paterson, director of the Philadelphia Mint, in which gold and iron were used, proved inconclusive. Better success was met with when experiments were made with the cooperation of Mr. Cornelius, a Philadelphia manufacturer of lamps. Their combined efforts resulted in the discovery that, under certain conditions of temperature, silver could penetrate copper. This was of especial interest to the manufacturer because certain effects in plated ware had perplexed him. He had never dreamed that one metal would "dissolve" another.

In the summer of 1844 the college calendar was revised so that Commencement was advanced from autumn to summer. The effect of this change was to abbreviate the faculty's vacation to a bare two weeks. The professor spent the entire time blowing soap-bubbles.

This orgy of bubble blowing was not evidence of eccentricity, however much it may have diverted his neighbors. It had its origin in a great national calamity. Moreover, to the student of Henry's work it marks the beginning of a series of logical steps which led him toward a study of molecular construction of matter and molecular forces.

On February 24, 1844, the U.S.S. *Princeton* latest and most heavily armed unit in the navy, steamed down the Potomac River to permit the President and a party of his guests to witness a demonstration of the vessel's special features, which included the first application of Ericsson's propeller to a ship of war, and two monster twelve-inch guns, the largest in the world. The climax to the outing was to be the firing of the great guns. After several rounds had been fired according to a prearranged schedule, the Secretary of the Navy requested that one more be fired for the entertainment of the guests. This time the gun exploded, causing frightful casualties among the gun crew and the civilian spectators. Mercifully, President Tyler escaped, but among the fatal casualties were his Secretaries of State and of the Navy, and several members of Congress.

The Franklin Institute appointed a committee to investigate the cause of the disaster, with Henry among its members. When bars of iron, similar to those from which the gun had been cast were submitted to breaking tests, Henry observed that they ruptured with a cup-shaped fracture. This led him to infer that the molecular cohesion within the interior of the metal was not the same as that near the surface. Apparently he was not familiar with Poisson's law, but his inference was sufficient to prompt him to make a further investigation into the molecular construction of matter.

The foregoing facts must be understood, otherwise the integration of a number of his subsequent papers will not be perceived. These papers might be mistaken as casual contributions to detached subjects, such as capillarity and the atomic constitution of matter, whereas they were different aspects of the same study.

The first communication evoked by the study of the bursting of the gun barrel was an extemporaneous address given before the members of the American Philosophical Society on April 5, 1844, and was continued on May 17.[57] It is of interest to observe that when this appeared in print it was immediately reprinted by Silliman in his *Journal* and in the *London and Edinburgh Philosophical Magazine* of June 1845.

In an opening passage of his remarks we find a clue to the greater mystery which was occupying Henry's mind, although his remarks, when printed, bore the title of "On the Cohesion of Liquids." He remarked: "Very erroneous ideas are given as to the constitution of matter in the ordinary books on Natural Philosophy."

He explained that he had measured the tenacity of a soap film by "weighing the quantity of water which adhered to a bubble before it burst." The thickness of the film was determined from observation of its color in accordance with Newton's scale of thin plates. He concluded that the molecular attraction of water for water was not, as Young suggested in his theory of capillarity, 53 grains to the square inch, but several hundred pounds, and equal to that of ice for ice. The strength of the soap film, he declared, was not due to an increase in molecular attraction, but to a reduction in the mobility of the molecules.

As he pursued his experiments he learned that the cohesion of unadulterated water was greater than that of soapy water. He explained that the change from the solid to the liquid states was not due to the destruction of cohesion, but to the neutralization of "the polarity of

[57] Amer. Phil. Soc. *Proc.* Vol. IV, pp. 56, 84; *S.W.* Vol. I, p. 217.

the molecules so as to give them perfect freedom of movement around every imaginable axis." [58]

It should be noted that in his investigations of capillarity, tensile strength, and related phenomena, Henry followed much the same path as that taken by Joseph Plateau of Ghent, who is much better known in this field. The latter, whose *Statique Expérimentale et Théoretique des Liquides* was not published until 1873, used the same type of wire rings and discs that Henry used to induce free forms on the surface of the films to assume new forms depending on the equilibrium of the surface tension. Plateau extended the investigation in his experiments with films of oil on water.

We have regarded Henry as solely an experimental physicist, and we can discern how, when occasion required, he was able to direct his experiments toward a practical end. This aspect of his mental endowment was to receive sharper definition in his later work, but before this stage was reached he had a theoretic phase, and we perceive that his brain has grown in solid depth.

On December 20, 1844, he made an oral communication to his favorite audience on the "Origin and Classification of the Natural Motors." [59] At first glance it seems that this was a clean break from the studies on capillarity, tensile strength, and so forth. This view is erroneous. His mind was slowly working toward the conceptions of the correlation of forces and the principle of the conservation of energy, one of the main currents of nineteenth-century science. This conception was too bold to be attained in a single stride. In a succession of papers he treated individual aspects of the superior subject as they occurred to him.

The germ of the thought might have been implanted in his mind by reading an anonymous treatise entitled *Heat, Light, Electricity, and Magnetism* (Cambridge, 1827) which was added to the Albany Institute library while he was librarian. This book might conceivably have been inspired by another which presented a confused suggestion of the identity of these phenomena, written by the London lawyer C. C. Bompas, and called *Essay on the Nature of Heat, Light, and Electricity*. This book, published in London in 1817, was in the library of The Franklin Institute in Philadelphia, to which Henry had ac-

[58] *S.W.* Vol. I, p. 217.
[59] Amer. Phil. Soc. *Proc.* Vol. IV, p. 127; *S.W.* Vol. I, p. 220.

cess. The idea could have come to him from either source, but Henry's development of the doctrine is based upon more plausible reasons than either of these works had to offer.

Henry had to struggle to give form and cohesion to a cloud of nebulous and abstruse thoughts vaguely related to the conservation of matter, the conservation of forces, and chemical combination. The first clear conception of a fundamental principle was his suggestion that the correlation of forces could be extended to embrace the living machine, and his first effort at clarification was the classification of the natural motors.

We need not express surprise that scientists were striving to find universal forces which would eliminate the confusion presented by numerous individual powers to affect physical conditions. We have only to turn to the influential contemporary school of New England writers who were engaged in propounding a gospel of transcendental metaphysics which, in its way, was the literary equivalent for the universality of physical forces.

The series of papers which Henry contributed to this subject points to his apprehension of the universe as a unity controlled by a few elementary laws. There was no originality in this conception. It had been expressed by both Emerson and Carlyle. What was mystical and religious to them, Henry hoped to provide with a basis of positive knowledge.

The doctrine which we now know as the conservation of energy has a long history of gradual growth. The experiments of Rumford, Davy, and Carnot had suggested the equivalence of mechanical and thermal energy. The marvellous experiments made possible by the discovery of Voltaic electricity had permitted Davy, Ampère, Oersted, Faraday, Melloni, Joule, and Henry to bring to light facts which tended to show that the so-called forces at work in heat, light, electricity, and magnetism were related. It had been demonstrated that each of the "natural forces" could be created at the expense of another. From this sprang the idea of the transformation of forces.

However, the conception was nebulous. The second edition of Whewell's penetrating *Philosophy of the Inductive Sciences,* issued in 1847, contained no mention of this idea, nor did it make any allusion to the general law of conservation. It does not even mention Joule's announcement in 1843 of experiments which enabled him to form a correct view of the mechanical equivalence of heat.

In his first address on the natural motors Henry said that the train of thought had been suggested by some remarks of Mr. Babbage "in his work on the economy of machinery,[60] and to the later researches of the German and French chemists [Liebig, Dumas, and Boussingault principally] on the subject of vital chemistry." Henry declared that no one had attempted to correlate these ideas, so that he cannot have read the first brief but brilliant paper of Julius Mayer, published in 1842.[61]

He divided his "motors" into two classes, those which were referable to celestial disturbance—water, tide, and wind power; and those which were "referable to that which is called vital, or organic action" —animal power, steam, and other powers developed from the combustion of nature's fuels.

The report of Henry's remarks is brief, but it carries undertones of significance. He is quoted as saying:

> These natural motive principles are not always employed directly in producing work, but are sometimes used to develop other powers by disturbing the natural equilibrium of other forces, and in this way they give rise to a class of mechanical motors which may be called intermediate powers. It will be evident on a little reflection that the forces of gravity, cohesion, and chemical attraction, with those of the "imponderable" agents of nature, so far as they belong to the earth, all tend to produce a state of stable equilibrium at the surface of our planet.
>
> ... It will be found that while the approximate source of every power is the force exerted by matter in its passage from an unstable to a stable state of equilibrium, yet in all cases it may be referred beyond this to a force which disturbed a previous existing quiescence.[62]

Bearing in mind that our current system of nomenclature had not yet been developed, this passage would imply that Henry was in process of passing from a conception of the transformation of energy to the conception of its conservation. Was he aware that Descartes and Huygens had suggested that the total amount of kinetic energy

[60] *Economy of Machines and Manufacturers.* London, 1834.
[61] Bemerkungen über die Kräfte der umbelebten Natur. *Liebig Annal.* Vol. XLII, 1842, p. 233.
[62] Am. Phil. Soc. *Proc.* Vol. IV, p. 127; *S.W.* Vol. I, p. 221.

in a system was constant? Carnot's work on a mechanical theory of heat was not well-known, nor was Joule's study of the heating effects of electric currents (1843) available to Henry. Energy as an exact physical quantity was not clearly defined until 1847, and the word "energy" was not used scientifically until 1852. Henry proceeded warily, for he had made his reputation in the laboratory, rather than in the realm of theory.

Thus, according to Liebig, Dumas and Boussingault, the mechanical power exerted by animals is due to the passage of organized matter in the body from an unstable to a stable equilibrium; and as this matter is derived in an unstable stage from vegetables, and the elements of these again from the atmosphere, it would appear therefore to follow, that animal power is referable to the same sources as that from the combustion of fuels, namely the original force which separates the elements of the plants from their stable and original combination with the oxygen of the atmosphere. But what is this power which furnishes the plant with the material of its growth? Is it due to a constantly created vital power? Or, since its effects are never directly exhibited but in the presence of light, may not the opinion of many chemists of the present day be adopted, namely, that it is due to the decomposing energy of the sun's rays, which are found to exhibit a wonderful decomposing effect in cases where no vital phenomena are present?

If this hypothesis be adopted it must be supposed that vitality is that mysterious principle which propagates a form and arranges the atoms of organizable matter, while the power with which it operates, as well as that developed by the burning fuel and the moving animal, is a separate force, derived from the divellent power of the sunbeam. It is true that this is as yet little more than a hypothesis, and as such forms no part of positive science, but it appears to be founded on a clear physical analogy, and may therefore form the basis of definite philosophical research.

That phrase he uses: "it would appear to follow that animal power is referable to the same sources as that from the combustion of fuel," is noteworthy. It is the first explicit announcement, preceding Mayer's [63] by a twelvemonth, of a view now generally accepted.

[63] *Philosophical Magazine*, Vol. XXV, 1863.

Speaking of Henry's work in this field, Asa Gray, the great American botanist and Henry's close friend, declared: "If he had published in 1844, with some fulness, as he then wrought them out, his conception and his attractive illustrations, of the sources, transformation and equivalence of mechanical power, and given them fitting publicity, Henry's name would have been prominent among the pioneers and founders of the modern doctrine of the conservation of energy." [64]

As was the case with most of Henry's efforts to break down the partitions which divided science into watertight compartments, his work was overshadowed by the labors of others who classified, extended, or applied his primary deductions. His theory of the correlation of the physical and organic forces antedated the well known essay by Dr. William Carpenter, "On the Application of the Principle of Force to Physiology," [65] which covered the same ground.

Perhaps, however, the silence he adopted did not arise from lethargy or indifference, but was the natural course of a man who, having much to say, would speak only when the slumbering truth chose to spring fully to life.

Two years later, in November 1846, he approached the subject from another angle when he spoke to the same audience on the "Atomic Constitution of Matter." [66] After drawing attention to the antiquity of the atomic theory he made a somewhat slighting allusion to the limitations of the subject. The Philosophical Society reported: "Though at first sight speculations of this kind might belong exclusively to the province of the imagination, yet in reality he considered this hypothesis a fruitful source of valuable additions to our knowledge of the actual phenomena of the physical world."

One would have thought that the work of Berzelius would have interested him, but he does not appear to have been well-informed on it. Henry accepted the work of Davy, Dalton, and Berzelius with reservations. For him it was a working hypothesis. There were many scientists who were skeptical toward the atomic theory, and who preferred what they called "multiple proportions" as an explanation of the construction of matter. After all, no one had ever seen an atom.

When Henry took up the study of atomic physics it was not through deliberation, but because the subject had been thrust upon his notice

[64] Biographical memoir. *Memorial,* p. 64.
[65] See *Correlation and Conservation of Forces;* edited by Youmans, 1865.
[66] Amer. Phil. Soc. *Proc.* Vol. IV, p. 287; *S.W.* Vol. I, p. 255.

in an abruptly direct way by the bursting of a gun. A man of his type of thought could not very well have avoided it.

Near the beginning of his discussion, Henry said: "In constructing an hypothesis of the constitution of matter the simplest assumption, and indeed the only one founded on a proper physical analogy, is that the same laws of force and motion which govern the phenomena of the action of matter in masses pertains to the minutest atoms of these masses."

This seemingly logical approach, of applying Newton's laws to atomic particles, had given the atomists great trouble, and no one had been able to devise a suitable mechanical model which would both obey the laws of motion and yet exhibit the other properties which the atomic theory had been advanced to explain. Henry commented: "This hypothesis readily explains the statical properties of bodies, such as elasticity, porosity, impenetrability, solidity, liquidity, crystallization, resistance to compression when a force is applied to either side of the body, etc.; but it fails to account for the dynamic phenomena of *masses* of matter, or those which are referable to the three laws of motion."

The eighteenth century Jesuit priest Boscovitch had postulated that atoms were mere centers of attractive and repulsive forces acting at a distance, and that there was actually no collision between atoms. Ampère, Gauchy, and Faraday were among those who regarded atoms as unextended, or as simple centers of force. Henry could not agree.

"A portion of matter," he contended, "consists of an assemblage of indivisible and indestructible atoms endowed with attracting and repelling forces, and with the property of obedience to the three laws of motion."

It becomes clearer as Henry progresses that he was experiencing difficulties in reconciling the logical deductions from his own experiences with preconceived notions. He encounters difficulties with the corpuscular theory of light, and tried to bring it into harmony with a wave theory. "If we adopt the theory of undulation, the phenomena of the 'imponderables' (as they are called) are merely the results of the motions of the atoms of the ætherial medium combined in some cases with the motion of the atoms of the body."

He concluded his remarks with the admonition that he has been treating with the fundamental conception of the atomic construction of matter, and not the "arrangement of the atoms into systems or groups, which are necessary to represent the varied and complicated mechanical and chemical phenomena exhibited in the physical changes going on around us. Though he could not at this time attempt to give any details of application of this hypothesis, he drew attention to one class of facts of which it is important to furnish an expression in the arrangement of the atoms. He alluded to the facts of polarity, or those which exhibit the action of opposite forces at the extremities of molecules or of masses. The north and south poles of two magnets, brought together, neutralize each other; the attraction of one is balanced by the repulsion of the other; and the point of junction is without action on a third ferruginous body. In the same manner apparently, two chemical elements which enter into combination exhibit a neutralizing effect, which indicates the existence of polar forces in the phenomena of chemical action."

Thus we perceive that by the association of electric polarity with chemical affinity, Henry again appears in the role of pathfinder, although he was only independently arriving at the conclusion which Berzelius had reached when he affirmed that all atoms contained charges of either positive or negative electricity. Their work was fruitless; both men were too far in advance of their times.

Some of Henry's papers had been reprinted in Silliman's *Journal* and in the British *Philosophical Magazine,* but this address did not arouse the interest of the editors, and they failed to encourage Henry to expand his views on the subject, or even to further diffuse the ideas he had expressed.

The series of papers which are here grouped together regardless of chronology have a unity. Together they resemble the movements of a symphony. A certain theme runs through each, played with variations and elaborations, each complete in itself, but all necessary to each other to finish the grander conception.

Five years were to pass before Henry returned to the main theme, but, when he did so, in 1851, he declared that he had in mind the subject of his 1846 address. At a meeting of the American Association for the Advancement of Science on August 21, 1851, he presented a paper on "The Theory of the So-called Imponderables," [67] which shows he had taken an appreciable stride toward the fuller perception of the conservation of energy.

[67] A.A.A.S. *Proc.* Vol. VI, p. 84; *S.W.* Vol. I, p. 297.

The "imponderable fluids," weightless and incompressible, were an inheritance from early science. Light, caloric, phlogiston, electricity, and magnetism presented insurmountable obstacles to the comprehension of those men who had to rely upon sensory perceptions for measurement, and whose scientific horizon was bounded by Newton's laws.

From his objective consideration of the motions of the planets in deducing the law of gravitation, Newton arrived at conclusions which, by sheer weight of his authority, moulded the thought of scientists for generations. Most influential of the conclusions derived, for our present purpose, was the conception of matter as being constructed of tiny solid corpuscles in the universal void. In order to provide a frame of reference in the void, Newton was led to postulate the ether, which was accepted as an imponderable.

Henry was a physicist in the Newtonian tradition, but it is quite evident in this paper that he was having difficulty in assimilating the classic view of the imponderables. He seemed to derive a certain amount of satisfaction from showing that if Newton's theory did not precisely fit, it was nevertheless a useful working hypothesis in explaining a unity of force which would account for light, heat, electricity, and magnetism.

> The discoveries of the last few years, he said, have tended more and more to show the intimate connection of all the phenomena of the "imponderables"; and indeed we cannot avoid the conclusion, forced upon us by legitimate analogy, that they all result from different actions of one all-pervading principle. Take, for illustration, the following example of the development of the several classes of the phenomena. An iron rod rapidly hammered becomes red hot, or in other words, emits light and heat. The same rod insulated by a non-conductor and struck with another non-conductor exhibits electrical attraction and repulsion. Again, if this rod be struck with a hammer while in a vertical position it becomes magnetic. We have here the evolution of the four classes of phenomena by a simple agitation of the atoms. We cannot, in accordance with the known simplicity of the operations of nature, for a moment imagine that these different results are to be referred to as many different and independent principles.

Next, Henry had to fit these phenomena into the conception of the Newtonian ether.

> If we refer all this phenomena to one elastic medium it will be necessary in order to explain the facts of electricity and magnetism, that we suppose this medium to be capable of accumulation or condensation in certain portions of space, and of being lessened in quantity or rarefied in other portions; also that in its return to the normal condition an actual transfer of the medium takes place. It follows from these assumptions that the fluid withdrawn from one portion of space must leave an equivalent deficiency in another, or in other words, that the amount of positive action must be equal in all cases to that of the negative. Further, since it appears from observation that the ætherial medium can only be condensed or accumulated in certain places by the insulating powers of ordinary matter, no electrical phenomena can be exhibited except in connection with such matter; hence electrical action cannot be expected in the regions of celestial space.[68]

Thus we see how the great scientific ideas of the age, the conservation of energy and the atomic constitution of matter, developed in Henry's mind. He was firmly attached to the Newtonian tradition, even in his consideration of electromagnetism, where it was difficult to apply and with which he was most familiar. At the end of his essay on the imponderables he said:

> The most difficult phenomena for which to invent a plausible mechanical explanation, connected with this subject, are those of the attraction of the two wires transmitting a current of electricity, and the transverse action of a galvanic wire on a magnetic needle. The theory of Ampère, though an admirable expression of a generalization of the phenomena of electromagnetism, is wanting in that strict analogy with known mechanical actions which is desirable in a theory intended to explain phenomena of this kind.

[68] *S.W.* Vol. I, p. 304.

Nevertheless, bound though he was to Newtonian ideas, he well understood the function of radical hypotheses in the advancement of science. Speaking of the atomic theory, he said

> that the legitimate use of speculations of this kind is not to furnish plausible explanations of known phenomena, or to present old knowledge in a new and more imposing dress, but to serve the higher purpose of suggesting new experiments and new phenomena, and thus to assist in enlarging the bounds of science and extending the power of mind over matter; and unless the hypothesis can be employed in this way, however much ingenuity may have been expended in its construction, it can only be considered as a scientific romance worse than useless, since it tends to satisfy the mind with the semblance of truth, and thus to render truth itself less an object of desire.

No scientist today will disagree.

A much fuller and better-rounded presentation of Henry's views on the conservation of energy, with special relation to the organic world, was published in the Agricultural Section of the Report of the Patent Office for 1857, where it was securely buried from the notice of most scientific workers. Scarcely anyone but his more intimate friends read this enlargement of his views. However, when Mayer's work in Germany and Joule's in England began to be appreciated in this country, his friends urged him to disinter the article and to offer it to a wider audience, so that his work in this direction might receive the recognition it merited.

The outcome of this pressure was the reprinting of the article in the *American Journal of Science* in 1860, but it failed to win appreciation. None of the articles relating to the development of his views on the conservation of energy had been reprinted in the English journals. The lack of appreciation of the editors cannot be attributed to either personal or national prejudices. Men like Henry, Mayer, and Joule were too far in advance of contemporary thought to meet with a ready acceptance. Neither Henry nor Mayer received much attention, perhaps because they did not approach the matter experimentally. Joule did this, and did it so successfully in a series of experiments which his biographer [69] considers as technically the most difficult ever accomplished by a physicist, that his name is usually associated

---

[69] Reynolds, O. *Memoir of James Prescott Joule*. London, 1892.

with the doctrine of the conservation of energy. Yet Joule's first announcement before the British Association failed to arouse any favorable comment.

For the sake of his reputation it is to be regretted that Henry failed to give prompt and full accounts of his work. His reluctance to put his ideas into print until he was satisfied beyond any shadow of doubt that they were precisely correct in every detail also prevented him from being acclaimed as a pioneer in the great generalization of physics, the conservation of energy, as it had his discovery of electromagnetism. Before Mayer had carried his conception to completion (for his paper in 1842 was in a very condensed form), Henry had shown he was advancing steadily to a comprehension of the correlation of forces.

In praising Henry for his work along these lines it is not claimed that he paralleled the work of Joule, for the essence of Joule's labor was not so much to show that heat and mechanical work were interchangeable but that they were interchangeable at a fixed rate. Joule performed the experiments that clinched the matter, while Henry theorized. But when one recalls the controversy which raged in Europe upon the relative merits of Mayer, Joule, and the Danish engineer Colding upon priority of conception, one cannot refrain from adding Henry's name to those pioneers of the monumental conception.

Had his friend Bache, with a surer grasp of what was necessary to secure priority for a claim, been at hand to spur him into activity, Henry's efforts in this field might not be forgotten. But Bache had gone to Washington as head of the Geodetic Coast Survey. The trouble lay within Henry himself. He never courted publicity, not from fastidious dislike, nor from any disdain for applause, but simply because he never thought of it.

In following the development of Henry's thoughts on atomic physics and the transformation of forces, we have lost touch with chronology, and drifted far from his other work during the later years at Princeton.

He was a close observer of atmospheric electricity, and a thunderstorm would invariably find him with apparatus ready to study its effects. The first series of experiments he made on atmospheric induction with "two large kites, the lower end of the string of one being attached to the upper surface of the second kite, the string of each consisting of a fine wire, the terminal end of the whole being coiled around an insulating drum. . . . When they were elevated at a time when the sky was perfectly clear, sparks were drawn of surprising intensity and pungency, the electricity being supplied by the air, and the intensity being attributed to the induction of the long wire on itself." [70]

In another series of experiments with thunder clouds the metal roof of his home served as a condenser plate. In one of these experiments an electric discharge from the clouds was drawn between two chimneys of the house, and it was of sufficient "pungency" to raise the roof clear off the walls! [71]

Arising from his observations came two eminently practical papers in which he formulated a set of rules and directions for the protection of buildings against lightning,[72] the later one of which would be acceptable as a good safety engineering specification today.

In 1845 he began a series of observations on the heat radiation of sunspots. He was aided in this work by his brother-in-law, Stephen Alexander, who had withdrawn from the Theological Seminary and had joined the faculty of the college. The subject was suggested by the appearance of an unusually large sunspot situated in the center of the sun's disc when the investigation opened, and the reading of an article in a French journal in which the author, A. Gautier, speculated upon the influence of sunspots on terrestrial temperatures.[73]

Henry was also familiar with the work of Herschel, the English astronomer, who had sought to establish a correlation between the price of wheat and sunspot activity as furnishing proof that the solar spots were associated with greater emission of heat. Henry rejected his methods, dismissing the apparent analogy with the wise comment that too many causes contributed to variations in the price of wheat. Gautier based his investigations upon an examination of recorded temperatures at different parts of the world at the time of unusual sunspot activity. As the records were far from complete he concluded he did not have sufficient material on which to form a generalization,

[70] Letter to Rev. S. B. Dod. *Memorial*, p. 152.
[71] *Ibid.*, p. 154.
[72] On the effects of a thunderstorm. *S.W.* Vol. I, p. 193; On the protection of houses against lightning. *S.W.* Vol. I, p. 231.
[73] Gautier, A. Recherches relatives à l'influence que le nombre et la permanence des taches observées sur le disque du Soleil peuvent exercer sur les températures celestes. *Annales de Chimie*. Vol. XII, 1844, p. 57.

but he was inclined to believe that when most spots were active, or spots were very large, the terrestrial temperatures were lower.

Henry decided to restrict his investigations to the sun itself, thereby opening a new phase of solar physics.

A four-inch telescope was arranged to reproduce an image of the sun's disc on a screen in a darkened room. The unusually large sunspot which appeared on January 4 measured roughly two inches in diameter on the screen. The image of the sun's disc was then explored by means of a small thermopile. Cloudy weather interfered with observation and made the investigation less complete than the exacting Henry would have wished, but enough measurements were made to establish proof that the radiation from the spot was distinctly less than from "the surrounding parts of the luminous disc."

The results of the experiments were presented to the American Philosophical Society [74] and an account of the investigation was sent to Sir David Brewster, who judged Henry's letter to be of sufficient interest to read it to the members of the British Association.

Henry did not continue this line of research, perhaps because his term of original investigation was drawing to a close, but he did not drop it from his mind. Several years later, a young professor of mathematics at Georgetown College, came to consult him upon a physical problem. Henry was possessed of that happy faculty for infecting others with his enthusiasms, and when the young professor left the interview he was inspired by a desire to become an investigator of the heavens and to continue Henry's researches of the sunspot's temperatures. This young professor was P. A. Secchi, who later became astronomer and meteorologist of the Collegio Romano. While holding this post he announced the results of his classic contribution to the field of solar physics which had been opened to him by Professor Henry.

A great many other studies were conducted during these fruitful years at Princeton. He was, for example, still interested in the aurora and continued to investigate it, although not with much profit. He studied color blindness. He reflected heat from mirrors of ice, and obtained results which justified conclusions as to the source of the moon's reflected heat. One of his ingenious constructions was a thermal telescope consisting of a pasteboard tube, blackened inside and covered externally with gilt paper, by means of which he measured the heat

---

[74] Am. Phil. Soc. *Proc.* Vol. IV, p. 173; *S.W.* Vol. I, p. 224.

radiated by distant objects. He could detect heat emitted from a man a mile away, and that from a house five miles away. Turning his tube against the sky, he found the coolest part at the zenith.

He experimented on the flow of liquids under varying conditions; on sonorous flames passing through an eight-inch stove pipe; on the evaporation of water; and on heat and light. The extended notes he made on these and other subjects were destroyed by fire. Only his work on atmospheric electricity and meteorology, which covered a number of years, was spared from destruction. It was expanded and collected for publication.

Henry retained his early interest in meteorology on account of its practical importance, but also because of its relation to cosmic physics. In attempting to classify and correlate the data within his reach, he became impressed by the need for more extensive, continuous, and systematic observations than were being undertaken. During his visits to Philadelphia to attend meetings of the Philosophical Society he made the acquaintance of the famous meteorologists Loomis and Espey, who were conspicuous figures in the group of Franklin Institute members organized to study weather conditions in Pennsylvania. This was the best organized group of meteorologists in the country.

In 1838 a general plan based upon international cooperation for the observation of magnetic and meteorological changes was discussed by the British Association, and most European countries were persuaded to join in the scheme for gathering observations. As none of the American countries displayed an interest in the project, Henry and his friend Bache induced the influential American Philosophical Society to address a memorial to Congress urging the establishment of stations to make the necessary observations. The time was inauspicious, as the country was in the depths of a financial depression and Federal revenues showed a deficit. The government compromised by arranging for observations to be made at a few army posts.

In the ensuing years Henry collected a mass of notes on meteorological subjects—magnetic variations, auroras, whirlwinds, thunderstorms, etc. These were published in a series of articles in 1855–1859, which will be treated later.

FOUNDING THE SMITHSONIAN INSTITUTION

Henry had now reached his forty-ninth year. All the high promises that had accompanied him to Princeton had been fulfilled in the fourteen years of his duties there. He had found congenial companions, and duties well suited to his powers. He had been valued and honored by the members of the faculty, while the students held him in reverence. One of his students referred to him as one "who rose with the sun to instruct his pupils, eager after knowledge," and as "giving his heart and soul to the duties of the school."[1]

Yet, it was not solely as a college professor that he had won appreciation. He was accepted as the most prominent scientific worker in the country. At the prime of life, he was able to bring a mature judgement to the support of his mastery over the experimental method. Physical science had entered upon an epoch of astounding activity, and Henry held his position in the forefront of the onward march. By his later papers he had displayed a bold leadership in the ideas which were to be the nineteenth century's greatest contribution to science, the correlation of its inheritance, the establishing of the far-reaching concept of energy, and a departure from the materialism of the previous century.

He had made hundreds of experiments in electromagnetism and, according to his old student H. C. Cameron, he was working toward more interesting results. The same authority declares[2] that Henry

[1] Goode, G. B. The Secretaries. *Smithsonian Institution; 1846–1896*, p. 523.
[2] Cameron, H. C. Reminiscences. *Memorial*, p. 174.

said later that his duties prevented him from pursuing these investigations further.

It seemed that the ripest fruits of his investigations were yet to be harvested. Much was expected from him by his friends.

As for himself, his expectations were not excessive, nor much beyond the pleasures of the daily task. Now, at what may have been a climactic phase of work in physical science, he deliberately chose, to borrow his own expression, "to sacrifice fame to reputation," by voluntarily withdrawing from active work to become an organizer and administrator. This caused consternation among his friends who had no grounds for supposing he had any particular aptitude for this type of work.

The occasion which brought about the change was his acceptance of the post as Secretary of the Smithsonian Institution.

Henry's life divides naturally into three main periods. The first period revolves around his life in Albany, when he made his first experiments in electromagnetism and laid the foundation for his scientific fame. The second period was that passed at Princeton, during which he continued his scientific investigations and made himself the foremost scientist in America. At the conclusion of this period he took what was probably the most important decision in his life in order to embark upon his final major task. The decision he reached excluded all personal and emotional factors. With complete philosophical detachment he evaluated the needs of American science and threw in his lot with the Smithsonian Institution to promote the satisfaction of those needs.

Hitherto his life had been devoid of dramatic situations. Its interest lay in the part Henry had played in the advance of human knowledge, and it changed only because he moved on to a wider plane of influence.

It is a mistake to suppose he was exhausted scientifically because he no longer devoted his energies to fresh discoveries. He had made valuable contributions to science, and there was every indication that he could continue to do so. Yet he felt his place was no longer in a laboratory, but rather in the van of the nation's science, putting at the disposal of his people and all humanity his gifts as a worker and a thinker.

Not that he felt he was forsaking experimental work. In 1850 he declared that "For the last three and a half years, all his time and all his thoughts had been given to the details of the business of the Smithsonian Institution. He had been obliged to withdraw himself entirely from scientific research, but he hoped that now the Institution had got under way, and that the Regents had allowed him some able assistants that he would be allowed, in part, at least, to return to his first love—the investigation of the phenomena of nature." [3]

The hope proved to be largely illusory. He never did regain the complete freedom of the scientist in the laboratory. He became the servant of the public, serving the needs of different government bureaus; and those needs, while varying enough in character, had one point in common—each one was of a severely practical nature. He made no further research into basic science during his years at the Smithsonian. All the research he performed was of direct economic and social value, a fact which should not have escaped attention in an age which judged achievement by practical results.

The practical nature of the research he was invited to undertake required wholly different qualities to be brought into play, different from anything Henry had as yet displayed, except by implication. These tasks, in which he had to deal with closely defined, concrete problems that permitted no delightful strayings into uncharted by-paths in search of concealed truths, were cheerfully executed and efficiently discharged. They merit greater attention than they have received. Henry moved from task to task, from investigation of deliberately selected hypothesis to drudgery imposed by external authority, so easily and so steadily, with such competence and assurance, that the mere volume of the work accomplished stands almost without equal by any other American scientist. Judged by any other standard it ranks equally high. And not the least valuable portion of it was performed in the public service to which we must now give attention.

Today the Smithsonian Institution is an acknowledged leader among scientific institutions. To hundreds of thousands of people who may be only slightly informed upon the nature or extent of its labors, it represents authority. For much of its high reputation, both among scientific workers and the lay public, Joseph Henry is directly responsible. "An institution is the lengthened shadow of one man," says Emerson, and the Smithsonian Institution still bears the ineffaceable marks of Henry's influence. Popular clamor could always

[3] A.A.A.S. *Proc.,* 1850, p. 377.

be stirred up over the seeming waste of public funds that offered no prospect of either immediate or direct return, but the wise provisions of the scheme for realizing the founder's wishes devised by Henry, and his firm control in holding to the aim, resulted in the establishment of an institution which soon acquired, and has long retained, public confidence.

Few institutions of learning have such a romantic origin. The founder of this essentially American organization was not an American; he had no associations whatsoever with this country. Without knowing something of his unhappy life it is not easy to understand why he should have made the United States his legatee.

James Smithson was the illegitimate son of Hugh Smithson, first Duke of Northumberland, and Elizabeth Macie, a lineal descendant of King Henry VII of England. He was, therefore, half brother of that Lord Percy who covered the retreat of the British troops from Concord. Born in 1765, he entered Oxford at the age of seventeen, and there acquired a taste for science which never left him. He was the best student of his year in chemistry and mineralogy.

Admitted a Fellow of the Royal Society in 1787, he read a number of papers before that body. He was also the occasional contributor to the scientific journals of the day. The bulk of his manuscripts were lost in the destructive fire at the Smithsonian Institution, but William Rhees painstakingly traced the few which were published.[4] These show Smithson to have been an assiduous student of chemistry who achieved a modicum of success with primitive apparatus. He did not add to the theory of science; but a striking feature of his writings, as compared with those of his contemporaries, is the modernity of his views. He certainly helped to enlarge knowledge of the mineral species, one of which he identified, and to which his name was given.

Smithson was on intimate terms with some of the most notable scientists of his age, men like Wollaston, Cavendish, and Arago. Rejected by English society because of the bar sinister on his coat-of-arms, he established a home at 121 Rue Montmartre in Paris. This became a famous meeting place to which the more interesting American visitors to the French capital found their way.

In trying to uncover a motive for leaving a fortune to found a scientific institution in a land he had never visited, we must bear in mind that, in spite of his place in the scientific community and the

[4] Smithsonian Miscellaneous Collections, 1879. Vol. XXI.

sincere affection he gained from numerous friends, he was socially a disappointed man. This was undoubtedly a smart to his vanity, for he longed for social recognition. In a fit of pique, he boasted: "The best blood in England flows in my veins; on my father's side I am a Northumberland, and on my mother's side I am related to kings, but this avails me not. My name shall live in the memory of man when the titles of the Northumberlands and the Percys are extinct and forgotten."

It was a prophetic boast. The titles he mentioned are neither extinct nor forgotten, but it is undeniable that his name is better known.

Deprived of the station in life which he believed to be his by right of parentage, Smithson was worthy of admiration for his strivings to acquire a name in his own right. But in spite of prolific writings on literature, history, science, and the arts he failed to carve with his pen a deep niche in the memories of his fellow men. He had, perhaps, too high an opinion of his own abilities, for relatively few of his writings found their way into print.

He renounced his claims to fame with reluctance, and it was not until he had passed the age of sixty years that he caught a glimpse of the manner in which he could gain celebrity. He made a will in which his fortune was left to a nephew, with a provision for its reversion in case his nephew died without heirs, "to the United States of America, to found at Washington, under the name of the Smithsonian Institution an establishment for the increase and diffusion of knowledge among men."

The perplexing factor in the provision of this will is the choice of Washington as the site for the projected Institution. Most writers have been satisfied to ascribe Smithson's choice to the disappointment he felt over the humiliations he suffered through the repulsions of English society. This view overlooks the fact that he had long harbored an intention of bequeathing his fortune to the Royal Society. Even admitting the sense of injury he nourished at rejection by his fellow-countrymen, why should he not have chosen Paris, where he had found friends and happiness among a people who held no grudge against a man suffering through no fault of his own?

We cannot offer an authoritative answer to this question, but we do venture upon a suggestion.

At the time when Smithson made his will, several Americans were in Paris; artists like Dunlap and West, novelists like Irving and

Cooper were among the figures who visited the city. Among the vagrant Americans at the time were the oddly-assorted pair of friends Tom Paine and Joel Barlow. They were precisely the type of men who would present their letters of introduction at the Rue Montmartre.

Barlow had a vision of making Washington the center of intellectual life for the whole world. He had worked out a plan incorporating the establishment of a national university, a school of mines, an art school, a library, a museum, an observatory, and a mint. There was not much that art or science could provide for which Barlow failed to account in his program. The conspicuous omission was an institution that would correlate the activities of the others and publicize their achievements.

Barlow was eloquent over his vision of an intellectual center in the New World. We know that he met Smithson on his travels in Europe, and we may well believe that the English expatriate got a glowing account of the Athens that was to rise on the banks of the Potomac. He may have painted the picture of the intellectual paradise in such convincing poetry that he aroused in Smithson's mind the expectation of undying remembrance by association with it.

He could not hope, by leaving his fortune in London or Paris, that any institution bearing his name could outshine the Royal Society or the French Academy, but there was an excellent chance that the Smithsonian Institution would achieve permanent renown by taking the lead in a new country.

The Smithsonian bequest to the American nation was received with only a qualified gratitude. When President Jackson notified Congress of it, and a judiciary committee recommended acceptance, the bequest instantly aroused fears in certain political quarters. Foes of the President argued that Congress lacked the authority to accept such a gift. However, it became obvious that this argument stemmed from an antipathy to any measure which hinted, however remotely, at strengthening a centralized government. Smithson's scheme was on a plane which the pedestrian and political mind could not comprehend, and it was to encounter many obstacles before a settlement was reached.

Through the good will of the English law officers the claim was "rushed" through the Chancery Division in two years. That this was a genuine concession will be appreciated by readers of Dickens' *Little*

Dorrit, where the delays of the "Circumlocution Office" are described. Finally, the legacy in the form of 105,960 golden sovereigns and "eight shillings and seven pence wrapped in paper" was delivered in New York. No other sum of half a million dollars can ever have awakened such a widespread lust for its possession or control.

The bequest specifically set forth that the money was to be used "for the increase and diffusion of knowledge among men." The language is plain enough, yet those nine words were found to conceal so many pitfalls that Congress floundered about for nine years without finding a solution to the problem they posed.

The fund, once regarded as an unconstitutional gift, suddenly became eminently desirable. John Quincy Adams pleaded that it be devoted to the erection of a grand observatory; others wanted it for a national library or a national university; colleges pleaded their needs for a share; various libraries needed a portion or the whole of the sum; societies of all kinds competed for the money in order to defray the costs of printing their transactions; advocates of an agricultural college affirmed their rights to the entire legacy; while one eloquent group proclaimed they would not be content with anything but an endless series of lectures.

The narrowest escape the bequest ever had was when it nearly fell into the hands of the National Institution, founded in 1841, largely through the efforts of Joel R. Poinsett of South Carolina. The avowed intentions of its members (Henry among them) was to assemble scientific collections, and its leaders cherished the hope of obtaining the Smithson funds to attain this object. Just in the nick of time it was pointed out that while these objects were highly praiseworthy, the National Institution was a private corporation and, as such, could not inherit a bequest to the American people.

Finally, Congress framed an Act establishing the Institution. Among its provisions were the obligations to support a library, a museum, and an art gallery. Adams had renounced his observatory, but he was especially worthy of commendation for the earnestness with which he urged that the capital of the fund be kept intact and that the activities of the Institution be restricted to those which could be supported from the annual income. This was a feature of the scheme which Henry was to defend with true Scottish frugality.

Robert Dale Owen, who was active in the idealistic experiment in living in New Harmony, Indiana, was also prominent in defense of the fund against predatory attacks by well-meaning but mistaken zealots who, in the pursuit of their individual schemes, would have depleted Smithson's principle beyond repair.

As Chairman of the Organizing Committee, Owen and Henry's old friend Bache performed excellent service in extricating Smithson's essential idea from the tangle of conflicting plans which bedevilled Congress. Largely through their advocacy, a strong Board of Regents was appointed to govern the Institution. Both Owen and Bache were appointed to this Board.

Once the Regents were appointed they lost no time in approaching Henry, as the foremost American scientist, to obtain his views upon what the aims and objects of the Smithsonian Institution ought to be.

This proved to be a happy choice. Henry had given a great deal of thought to the disposal of the Smithson legacy from the moment when he first heard of it during his visit to England. Having taken the trouble to acquaint himself with the circumstances of Smithson's life, he had learned that the earlier will had provided for the fortune going to the Royal Society, but that the provision had been scratched out of the will in anger when the Society had refused to publish one of its would-be benefactor's scientific papers. This original choice of a beneficiary gave Henry a clue to the manner in which Smithson had intended his money to be used.

At the suggestion of the Regents, Henry wrote out his famous plan for "An Institution for the increase and diffusion of knowledge among men." Thus, by its very name, he drew attention to the first principles, which had escaped the attention of those who had taken part in the wordy discussions upon the character which should be forced upon the Institution.

One of the most important questions which Henry must have put to himself was: What part should the capital of a nation be expected to play in the advancement of scientific knowledge? The answer, under any circumstances, could be neither obvious nor easy, the terms depending upon the times and the greatness of the nation. Nor does it follow that the fundamental advances in scientific thought are necessarily associated with a capital city. But he concluded that the center of government, the focus of national life, would draw to itself the richest in national culture.

The consciousness of the inadequacy of the Smithson funds convinced Henry that the tasks undertaken by the Institution must be circumscribed or its entire mission must end in futility through the dispersion of its force. The riches of a national culture, as expressed through its literature, its arts, and its sciences, with their relations to education and industry, are far too extensive to come within the compass of a single organization. The limitations of the legacy made selection a necessity. The discretion exercised in making the selection furnished the chief merit to the document Henry prepared.

It is unnecessary to enter into the full details of the plan. A glance at its provisions must suffice, but the reader is recommended to study it in its entirety. The complete text appeared in the annual reports of the Regents for several years.

Knowing that Smithson was a writer who appreciated the meaning of the words chosen to express his thoughts, Henry based his plan upon a liberal interpretation of the clause "for the increase and diffusion of knowledge among men." To admit anything that lay beyond the provisions of those words was to him a violation of trust. Nor could the meaning of the terms be contracted to favor any one science or any group of men. He gave a certain latitude to the interpretation of the word "knowledge," but in regard to what it might mean in Smithson's mind as displayed in his published writings, he leaned heavily toward the sciences. Literature was eschewed, although language was admitted. Philosophy was avoided because it was incapable of proof.

To *increase* knowledge, Henry proposed the Institution should encourage men of talent to conduct original researches, and should devote a portion of its income to aid suitable workers to pursue research.

To *diffuse* knowledge, he proposed the publication of periodical reports on the progress made in the different branches of knowledge, and the occasional issue of treatises that might be of interest. Writings of note in foreign tongues should be made accessible to all students and, therefore, should be translated to reach the widest circle of readers.

Henry interpreted the words "among men" to embrace all mankind, not merely the residents of Washington, nor even the inhabitants of the United States. He thereby condemned by implication such adjuncts as a general library, a museum, and an art gallery, which were local in perspective. Henry did not disapprove of national institutions of this nature. He felt that if they were to be established they should be supported nationally instead of drawing their support from the Smithson fund.

His plan was the nearest approach conceivable to the founder's wishes as outlined in nine pregnant words. Needless to say it did not meet with widespread acclamation, for its sober provisions did not dazzle the popular imagination. But when the plan was submitted by the Regents to the judgement of the presidents of American colleges and learned societies, it received warm support. This approval did not suppress the clamor of those who advocated rival proposals, but it helped to advance the conviction that Henry's scheme most aptly fitted the founder's desires, and would confer the maximum benefit from a limited fund. Indeed, the benefaction was so small compared with the needs it sought to fulfill that only Henry's towering reputation carried the day against the supporters of narrower schemes.

An amusing story came to light regarding the draft of the plan which was to be submitted to the Regents. The final copy was in the handwriting of Henry C. Cameron, who became professor of Greek at Princeton, and who was a favorite pupil of Henry's. When the document was finished and handed to the author, Cameron studied the professor's face as he read. Suddenly, to the younger man's dismay, he observed a nervous twitching of Henry's lips, and the scrivener was informed with horror that a word had been omitted from the text. It was difficult, said Cameron when telling the story, to say whether he or Henry was most humiliated by the knowledge that the document intended for submission to the august body of Regents must be marred by a blemish.

After the Board of Regents were satisfied that Henry's plan provided the best solution to the problems posed in administering Smithson's bequest, they were confronted with the matter of finding an executive to whom authority could be delegated for carrying out the policy adopted. Alexander Dallas Bache, as a member of the original Board, is thought to have drafted the resolution passed by the Board of Regents which read: "It is essential for the advancement of the proper interest of the fund that the secretary of the Smithsonian Institution be a man possessing weight and character and a high degree of talent; and that it is further desirable that he possess eminent scientific and general acquirements; that he be a man capable of advancing science and promoting letters by original research and effort, well qualified to act as a respected channel of communication between the Institution and scientific and literary individuals and societies in this and foreign countries: and, in a word, a man worthy to represent before the world of science and letters the Institution over which the Board presides."

Was this definition of the secretary's qualifications deliberately written around Joseph Henry?

Certainly Henry was not active in seeking the appointment. At the time it was offered there was talk of his being invited to fill a chair at Harvard, an offer which undoubtedly would have been a strong temptation. Moreover, Torrey declared that Henry had been annoyed by a newspaper report that he was an applicant for the post at the Smithsonian.[5] He was neither applicant nor candidate. The appointment came as a surprise to Princeton and to him.

Bache had been active elsewhere. If he did some log-rolling on Henry's behalf it was only a return for his friend's efforts to secure for him the post of Superintendent of the Coast Survey.[6] Immediately after he had been appointed a Regent, Bache began to canvas the best-known scientists to secure their approval of Henry as secretary. Among those who responded warmly to the suggestion were Faraday, Arago, Brewster, Silliman, and Hare. On this occasion Sir David Brewster said that "the mantle of Franklin has descended upon the shoulders of young Henry," which would have been a perfect response to Bache's request for a testimonial if only young Henry had not been fifty years old.

The Regents agreed unanimously to offer the position to Henry. The acceptance of the offer having been received, his appointment was announced on December 3, 1846. On the following day, the *National Intelligencer* voiced the general opinion upon the appointment in this editorial comment:

> There has perhaps never been an occasion in the literary history of our country when so much depended on the decision of so small a number of men. The success of one of the most liberal institutions in the world, depends much upon the personal influence of the Secretary to be chosen by the Regents. Men of the

[5] Rodgers. *John Torrey,* p. 201.
[6] Odgers. *A. D. Bache,* p. 145.

highest literary distinction as well as personal merit were numbered among the candidates.

... Foremost among American savans stands the name of Franklin, a name which belongs to the world. Second, perhaps to Franklin only, stands the name of the philosopher of Princeton. If Franklin discovered the identity between lightning and electricity, Henry has gone further, and reduced electric and magnetic action to the same laws. It is impossible in a short compass to do justice to the beauty and simplicity of Henry's laws of the action of the imponderable agents. Whoever will read the progress of his discoveries as published in the Transactions of the American Philosophical Society will soon learn something of the spirit of inductive reasoning of which Henry's researches furnish one of the happiest illustrations. These discoveries are not confined in the sphere of their utility to the limited circulation of the volumes of that Society. The student of physical science may read the reprints of them and the enconiums pronounced upon them in every language of civilized man. ...

It is the man who gives dignity to the office, not the office to the man. We doubt not that the republic of letters throughout the United States will applaud the choice, and give to the Regents their cordial support.

Did Henry regret having to abandon the pleasant path of his collegiate existence to engage in executive duties?

There seems little reason to doubt that a continuation of this life would have conducted him to a higher pinnacle of fame in science, to an enlargement of the circle of congenial friends, and to a prolongation into the future of an existence that promised no riches, but which was free from financial worries.

Most of his Princeton friends opposed his going to Washington. They pointed out that he would be undertaking an occupation that was entirely foreign to his experience; he would be engaging in a career wherein he must grapple with discordant problems of finance, policy, and organization, and he might become so engrossed in these that he would have no time to make direct contributions to knowledge. They told him that he was the acknowledged leader in the field of scientific investigation whereas, in this new work, he would become submerged in an institution. The new life might easily become one without stimuli, a leaden routine of duties.

Henry displayed no great enthusiasm about accepting the new post. The obligation to carry out the unwise provisions of the Congressional Act must have hung depressingly over him. One can imagine the uncertainties that must have tormented him at the greatest crisis of his life. The solicitations of his friend Bache had much to do with his acceptance of the post. We have Henry's own testimony to this effect. In a memorial eulogy of Bache he said with regard to the latter's interest in the Smithsonian Institution: "It would be difficult for the secretary, however unwilling to intrude anything personal on this occasion, to forebear mentioning that it was entirely due to the persuasive influence of the professor that he was induced almost against his own better judgement, to leave the quiet pursuits of science and the congenial employment of college instruction to assume the laborious and responsible duties of the office to which, through the partiality of friendship, he had been called." [7]

He was under no illusions. To one friend he said he was "sacrificing future fame to present reputation." To another he wistfully recalled that his great idol, Newton, had made no discoveries after he became Warden of the Royal Mint.

Against this, Bache and his other friends of the American Philosophical Society pointed out the opportunities he would have to serve science on a wider scale. It was this type of argument which would convert him. He only had to be convinced that a course of action was the performance of a duty and he would follow it unswervingly. Although much of the service he could provide for science must of necessity be anonymous, that would not trouble him. He was never one to advance his own claims.

Moreover, he had some justification for supposing that he could continue his experimental work once the Institution was operating smoothly. The Regents recognized the loss to science which would be incurred by Henry's withdrawal from active work, and they proposed he should "continue his researches in physics, and to present such facts and principles as may be developed for publication in the Smithsonian Contributions."

Their intentions were of the best, but there was a general realiza-

[7] Nat. Acad. Sci. *Biographical Memoirs*. Vol. I.

tion that Congress had muddled the provisions of Henry's "Plan," and the wisest must have known that a bitter struggle lay ahead.

One consoling feature must be remarked. A line of retreat was left open. The authorities at Princeton generously assured him that, should his new duties prove to be too distasteful or too severe a test for his patience, his chair would be open to him, and he would find a warm welcome on his return.

The appointment as Secretary was made on December 3, 1845, and he assumed office on December 14. It was perhaps fortunate that he had been taught a creed which put little value on money. The salary of $3,500 a year, with an additional allowance of $500 until rooms were provided for him, while no equivalent of wealth was an assurance against actual want, but hardly suitable recompense for the duties he was expected to fulfill.

From the outset the obstacles to success loomed larger and solider than they had when viewed from Princeton. On the day after Christmas he wrote a letter to President Nott of Union College reflecting the doubts which had crept into his mind.

> Please accept my thanks, he wrote, for your kind congratulations on my appointment to the office of Secretary of the Smithsonian Institution. I am not sure, however, that my appointment will prove a subject for congratulation. The office is one which I have by no means coveted and which I accepted at the earnest solicitation of some friends of science in our country, to prevent its falling into worse hands, and with the hope of saving the noble bequest of Smithson from being squandered on chimerical or unworthy projects. My first object is to urge on the Regents the adoption of a simple practical plan of carrying out the design of the testator, viz: "the increase and diffusion of knowledge among men."
>
> For this purpose in my opinion the organization of the Institution should be such as to stimulate original research in all branches of knowledge, in every part of our country and throughout the world, and also to provide the means of diffusing at stated periods an account of the progress of general knowledge compiled from Journals of all languages.
>
> To establish such an organization, I must endeavor to prevent expenditures of a large portion of the funds of the Smithsonian

The Smithsonian Institution, as seen from the northwest after the fire of 1865 which destroyed a portion of the upper part of the building.

THE ELECTRO-MAGNETIC TELEGRAPH.

A DEFENCE

AGAINST THE INJURIOUS DEDUCTIONS DRAWN FROM THE

DEPOSITION OF PROF. JOSEPH HENRY

(IN THE SEVERAL TELEGRAPH SUITS),

WITH A CRITICAL REVIEW OF SAID DEPOSITION, AND AN EXAMI-
NATION OF PROF. HENRY'S ALLEGED DISCOVERIES,
BEARING UPON THE ELECTRO-MAG-
NETIC TELEGRAPH.

BY SAMUEL F. B. MORSE, LL.D.,

PROFESSOR IN THE NEW YORK CITY UNIVERSITY, &C., &C., &C.

INTRODUCTION.

THE deposition of Prof. Joseph Henry (substantially the same in each of four Telegraph suits) has been erected by the ingenuity, and, I must add, by the sophistry of the counsel of my opponents into an apparently formidable barrier to my claims to originality in the invention of the Electro-Magnetic Telegraph.

While the facts he professes truthfully to state have been extravagantly dilated, and his position in relation to discoveries, alleged to have a bearing on the Electro-Magnetic Telegraph, immoderately exaggerated; yet the fault of this extravagance lies not altogether at their door. Prof. Henry has himself furnished by the ambiguity, as well as incorrectness of his narration, much of the specious material for their argument.

His deposition has been made to bear, not only on the legal points involved in my invention, but it contains imputations

The opening page of Morse's "Defence."

bequest on a pile of bricks and mortar, filled with objects of curiosity, intended for the embellishment of Washington, and the amusement of those who visit the city.

My object at present is to prevent the adoption of plans which may tend to embarrass the future usefulness of the Institution, and for this purpose I do not intend to make any appointment unless expressly directed to do so by the Regents, until the organization is definitely settled.

The income of the Institution is not sufficient to carry out a fourth of the plans mentioned in the Act of Congress, and contemplated in the reports of the Regents. For example to support the expense of the museum of the Exploring Expedition presented by government to the Smithsonian Institution, will require in interest on building and expense of attendance upward of 10,000 dollars annually. A corps of professors with necessary assistants will amount to 12,000 to 15,000 dollars. From these facts you will readily perceive that unless the Institution is started with great caution there is danger of absorbing all the income in a few objects, which in themselves may not be the best means of carrying out the designs of the Testator. I have elaborated a simple plan of organization which I intend to press with all my energy. If this is adopted I am confident the name of Smithson will become familiar to every part of the civilized world. If I cannot succeed in carrying out my plans—at least in a considerable degree, I shall withdraw from the Institution.[8]

There is neither elation nor pessimism in this letter, but only a frank admission that he was embarking upon his new duties with a knowledge that a struggle awaited him. He did not reconcile himself to the inevitable nor was he going to remain inactive. A duty had been assumed and he was determined to discharge it to the limit of his ability, but he would suffer no compromise with principles. How tenaciously he clung to the moral responsibility of carrying out Smithson's ideas was soon to be put to the test.

Shortly afterwards, when he was offered the opportunity to escape from Washington and to return to the laboratory, he repulsed it. Henry had barely settled in his secretarial office in Washington, his wife and family had not yet vacated their home in Princeton, when

---
[8] *Memorial*, p. 409.

an unique opportunity came to him to return to original research under well nigh ideal conditions.

Dr. Robert Hare, the professor of chemistry at the Medical School of the University of Pennsylvania, resigned his chair. The office he had occupied for so long was probably the most desirable scientific position the country had to offer. The financial conditions were highly favorable. The fees amounted to $5,000 or $6,000 a year while the duties consumed only half the year. Hare had been an intimate friend and great admirer of Henry, and it was his wish that the latter should succeed him.

As soon as Hare's chair was declared vacant it was offered to Henry, largely through the instrumentality of Fairman Rogers, the professor of civil engineering. Just as promptly, Henry declined it. The position was ideally suited to his tastes and his attainments. He might yet have scaled new and higher heights in the realm of science. Yet he decided he must finish his task in Washington.

After he had met the University's Board of Trustees and had announced his decision not to accept their offer, he went on to Princeton to rejoin his family. Crossing the college campus, he met young Henry Cameron, his former student. Setting down his carpet bag he unburdened himself to his young admirer. He told of the dilemma in which he had been placed by the invitation to fill a position in which he might have accomplished so much for himself and for science. But there was an insurmountable obstacle.

As he explained to Cameron, the situation was this: Had he accepted the offer of the chair at Pennsylvania, to which a handsome salary was attached, he exposed himself to the accusation of being attracted by pecuniary reasons and base motives! No wonder so many of his friends threw up their hands in despair at the man's unnatural conduct.

None the less, experimental science suffered a great loss when the professor decided that no taint of greed should be permitted to stain his unblemished reputation, as though anyone in his right senses would have accused this simple creature of being actuated by such motives. One feels that the situation in Philadelphia may not have been presented to him in the right light. It is unlikely that the University officials attempted to persuade him to accept their offer. Philadelphians of the Victorian era had the reputation of being a stiff-necked generation, unsympathetic toward an outsider who presumed to hesitate about accepting the opportunity of being admitted to a place of honor in their community. Still, it is strange that none of the members of the Philosophical Society interposed to plead with Henry, so that he might be persuaded even if he would have to condescend by accepting a higher remuneration for his labor.

This was not the only offer of its kind which came his way. So far, he had not encountered opposition to his plans; but later, when he was in the midst of discord in his duties, he was approached by John C. Calhoun of South Carolina, who offered him a chair in that state's university. Calhoun differed from the worthy Philadelphians; he did not hesitate to apply persuasion. A shrewd politician, he knew how to prepare and to present his case. He pointed out the sterility of the secretaryship as compared with the fruitfulness of a professorial chair, with the leisure it afforded for original research. Calhoun was in the Senate, fully aware of the forces at work to vitiate Henry's purpose, and he did not hesitate to point out the possibility, nay the probability, that the noble scheme for the Institution would crumble into dust through lack of appreciation of its merits. But he, too, met with a rebuff.

Henry acknowledged that the proffered position would yield him advantages for which the secretaryship could offer no compensation but, he declared, his personal benefit did not enter into consideration. He was "in honor committed to the Institution." When Henry gave this answer the master-orator of the South grasped him warmly by the hand, his dark eyes glowing with admiration as he said, "Professor Henry, you are a man after my own heart."

Allurements continued to be dangled before him. Twice, in 1853 and again in 1867, friends approached him with the request that he permit them to submit his name to the Board of Trustees for election to the presidency of Princeton, the college of his heart, where he had spent the happiest years of his life. These invitations must have tried his resolution to the utmost, but he was inflexible. For good or for ill, he had chosen to bind his fortune to the Institution and he was determined to support it with dignity and contentment to the end.

Science may have been deprived of a great investigator, a master of experimental research, but it had gained a great organizer, and American science was sorely in need of one who could organize and stimulate its efforts. Faithfully and unfalteringly he served science in another way than some of his friends wished. Too much fame had not dehumanized him. His gentle manner and sincerity made him approachable by the young and inexperienced; for his wisdom he was sought out by his fellow scientist. In that situation, he judged, he could be of greatest assistance.

And who will gainsay that he did not leave the nobler legacy to science?

Upon succeeding Henry in the position of secretary to the Smithsonian Institution, Professor Spencer F. Baird said: "There are numerous establishments in the United States, not of precisely similar character, but with the same general object, and with equal or larger funds of endowment, but which are scarcely known or even heard of outside the limit of the city in which they happen to be situated. The name of the Smithsonian Institution, on the contrary, is a familiar one in every part of the world; and it may almost be said that it is better understood, comprehended, and appreciated in the remotest parts of Europe than it is in some sections of the United States." [9]

Though this was the picture at the time of Henry's death, and as we see it today, it must be conceded that when the Professor first went to Washington this splendid vista was not discernible. The prospect was bleak and unpromising.

The year 1848 fell in a period of intellectual ferment and political unrest, with revolutionary movements in Europe and in America. The Mexican War had just closed, and by the treaty which ended the war the United States annexed a vast western territory. The discovery of gold in California and the conclusion of the postal convention between the United States and Great Britain were important factors in shaping economic events. Exploring expeditions were providing a geographic interest, and a notable engineering feat was the completion of the High Bridge aqueduct in New York city.

Yet the nation was spiritually unintegrated. Psycho-religious movements figured prominently in the news of the period, the two most notable examples being the first public demonstration of spiritualism and the vitalizing of Mormonism by the persecution at Nauvoo. During the next few years reforms and crazes were to flourish as never before; any fad that could feed upon the restlessness and superabundant nervous energy of the people was feverishly pursued.

The capital to which Henry came was little more than an archi-

[9] *Smithsonian Report for 1878*, p. 8.

tect's sketch of how a great city should look at its start. Ambitiously laid out over a wide area, the city struggled ineffectively to attain a unity for its widely spaced and unrelated buildings. Even after he had been a resident for twelve years, the capital failed to arouse praise. Henry Adams had a fling at it with his mordant pen. "As in 1800 and in 1850, so in 1860 the same rude colony was camped in the same forest, with the same unfinished Greek temples for workrooms and sloughs for roads."

The city was in simple truth a raw Southern town, with none of the traditions nor any of the picturesqueness but all the disorder, to which the porticoed Federal buildings failed to add grace. It was a strange site for a temple of learning that was to shed the light of knowledge among men.

If one could tolerate the long unhealthy summer it was possible to drowse through a little work, for the place was deserted and a man had no social distractions. Only when the houses of legislature were in session did the city wake up and life began to flow.

Robert Dale Owen, as chairman, and the members of the Building Committee of the Smithsonian Institution had already made progress with the building before Henry's arrival in Washington. The choice of a site was not at all propitious. A plot of ground on the Mall lying west of Seventh Street was selected. The location was remote from residential Washington, being situated on what was called the "Island," because it was isolated from the mainland by a strip of malodorous water called euphemistically a canal but which, in reality, was no more than an open sewer. The Island was reached by four bridges. The site of the Institution was waste land, unfenced, and neglected except when it was occasionally occupied by a camping regiment or a travelling showman.

It was such a desolate spot that, when the east wing of the building was completed and the first lectures were given, the Regents were compelled to improve the site by laying plank walks through the mud to enable visitors to reach the building.

Upon this site, Renwick's building was to be erected in white marble at a cost of $300,000. Before work was commenced the Building Committee changed their minds about the material and carried out the architect's idea of what he called a "Lombard" design in brown freestone. The mass of the building was relieved by a number of incongruous towers, the like of which are the despair of every museum curator and librarian, since they occupy valuable space and serve no known purpose. The towers were put to a use which the architect can hardly have contemplated but which helped to justify their existence. Miss Lucy Baird, daughter of Henry's assistant and successor as Secretary, told how "Many young naturalists who were studying in the museum as well as assisting in its work lodged in the Smithsonian towers. By the kindness of Professor Henry many of the unused rooms, too high up for business purposes and situated conveniently for access to their work were assigned to such young students as lodgings." [10]

Fortunately, the grounds were laid out in good taste by Andrew Downing.

Henry thoroughly disapproved of the whole affair. Apart from the unsuitable design, he lamented the expenditure of the funds upon a pile of bricks and mortar, however imposing the result might be. The knowledge that the accumulated interest upon the funds of the bequest were almost sufficient to defray the cost in no way mitigated his displeasure. Had he been left to his own devices he would probably have rented a second floor of some premises on a quiet street to accommodate his staff. But Congress had willed the great structure on the Mall and it went up slowly.

In his opposition to the building Henry may have been in the wrong. He was either insensitive to or ignored the psychological influence which a large impressive building exercises upon the minds of men. Today, his opinions would be mocked. He never ceased to deplore the elaboration of the structure in which he had to work and the strain of its maintenance upon the funds he had to administer.

The cornerstone was laid on May 1, 1847, to the accompaniment of benedictions, brass bands, and artillery salvoes, in the presence of six thousand spectators. It was the Building Committee's pious wish, happily not fulfilled, that the edifice should be an example of and stimulation for domestic architecture of an essentially American character.

Provision was to be made in the building for the display of objects relating to natural history, a geological and mineralogical "cabinet," a chemical laboratory, a library, an art gallery, and lecture rooms.

Henry was firmly opposed to most of these departments. The building having been started and its functions defined by Act of

[10] Dall. *S. F. Baird*, p. 208.

FOUNDING THE SMITHSONIAN

Congress, he could not actively obstruct them, but as his letter to President Nott of Union College made clear, he would not continue in his position unless he could persuade the Regents to modify some of the provisions and revert to his own simpler plan.

He did his utmost without subterfuge or guile to repair the injury done to the Institution by a well-meaning Congress. This was bound to lead to friction. He knew he would have opponents, but from the day he assumed office he labored without remission to lighten the burden on the finances caused by those departments he regarded as having been inflicted upon his scheme by the erroneous interpretation of Smithson's bequest. Only by the exercise of unlimited perseverance and patience was he able to bring the Institution within the framework he had planned.

One would like to think that Henry's policy found unanimous support among the scientists in whose benefit it was devised, but this was not the case. The naturalists especially were displeased. They wanted specimens rather than information about them, and they could not understand why Henry did not wish to undertake these collections of specimens. There must have been many struggles to win over the Regents to this phase of his policy. Indeed, he must have threatened to resign at an early date rather than sacrifice his principle of putting a check on expenditures in order to preserve the capital sum of the bequest. As early as June of 1847, Bache was pleading with him not to contemplate resignation.[11] In a letter to Baird, who was awaiting appointment as his assistant, dated December 11, 1849, Henry alludes to the subject again, writing that the Regents wished "and still wish to be able to state to the public that after completing the building and getting the Institution under way, they have added 150,000 dollars to the original fund. Every proposition which interferes with this plan is received with coolness. The plan was originally suggested by the probability that I was about to resign the office of Secretary." [12]

He was adamantine in his determination to carry out his plan of organization and to resist any encroachment upon it. "You know," he wrote in another letter to Baird, "that I accepted my position with the understanding that I should be allowed to carry out my plans of active operations, and that in accordance with this understanding I

[11] Odgers. *A. D. Bache*, p. 166.
[12] Dall. *S. F. Baird*, p. 199.

refused to accept a position much more in accordance with my taste as well as my pecuniary interest. I now find myself, however, very much restricted by the compromise of the Boards and the diminution of our income. I am therefore the more determined to guard myself and what I deem the interest of the Institution from further restrictions in the carrying out of my plans." [13]

Hampered in the attainment of his objects, Henry applied himself to the task of winning a majority of the Regents around to the execution of his original plan. He was conscious that several members of Congress and many ordinary citizens were desirous of having something more tangible and showy than the knowledge he proposed to encourage and diffuse.

The public was willing enough to support schemes for surveying new territories, for promoting agriculture, even for encouraging geological surveys. From all of these a sound practical return was to be expected. An institution having for its aims the promotion of pure scientific knowledge could not expect much support from the multitude who placed their faith in utilitarian developments. Henry had to proceed warily.

Because of his earlier work in electromagnetism, it might be supposed that Henry would show discrimination in favor of physical studies in the work of the Smithsonian, but this did not influence his actions in the least. His beginning showed he was approaching the work with a wide and catholic spirit.

He placed great faith in the printing press as an agent for diffusing knowledge, but he knew that his ideas upon its use would run counter to the public's wishes. The popular conception of advancing knowledge by means of publications was to encourage the issue of tracts like the *Penny Magazine* or other elementary books. He had been brought to a realization of the beauty and the fascination of the laws of nature by just such a book, but Henry had no intention of yielding to the demand for such literature.

He knew that private publishers could be depended upon to issue at low prices all the works of this kind the demand could absorb. What he sought to encourage was the publication of learned memoirs, often voluminous and elaborately illustrated, which did not appeal to the private publishers because of the restricted market and the high cost of production. In his first report to the Regents he announced

[13] *Ibid.*, p. 208.

that the Institution would sponsor the publication of *The Ancient Monuments of the Mississippi Valley,* the classic work of Squier and Davis.

Before 1847 the unaided work of a few pioneers like Caleb Atwater had called attention to the antiquities of the country, but public interest in them was not aroused until the Smithsonian Institution quickened the investigation of the mound builders' work through this and later publications. The favorable impression created by the appearance of this work by Squier and Davis convinced Henry that the publication of major scientific books was one way in which the founder's wishes might be fulfilled.

He instituted two types of publications. One was the "Smithsonian Contributions to Knowledge," quarto volumes of original studies, either the result of special investigations authorized and directed by the officers of the Institution or carried out under other auspices and submitted for publication. Henry set a very high standard for these works, and because the standards have since been rigidly maintained, the volumes in the series are prized among scientific workers.

The second class of works, called the "Smithsonian Miscellaneous Collections," were octavo volumes of descriptive material in natural history, physical tables, reports on the progress of science, and the like.

At the time of Henry's death, twenty-one volumes of the "Contributions" and fifteen volumes of the "Collections" had been issued and distributed freely throughout the world. There was no coddling of the dilettante. These publications were not intended for the general reader, but for earnest students. By the distribution of the papers in the "Miscellaneous Collections," scholars in all parts of the world were given access to these valuable papers.

Henry's share in the preparation of these volumes for publication did not terminate when a decision had been reached to publish them. He was fully alive to the responsibility of editorship, for he had received the confidence of Silliman, who had exercised herculean efforts to keep alive the *American Journal of Science* and to maintain a high standard for the material he published. Henry did not have to contend with the formidable financial ordeals through which Silliman steered his journal, but he shared the older man's sense of responsibility for the printed words which he was instrumental in presenting to his readers. How gravely he regarded the task of editorship and what attention he devoted to its details may be gathered from the review of Silliman's editorial duties which he wrote at the time of his friend's death.[14]

He gave each work his closest attention, editing it so that obscurities might be cleared, pruning over-rich metaphors, excising hasty or inexact generalizations, and correcting inaccuracies. Curiously enough, yet characteristically, he never used the "Collections" for any of his own papers, although some could have been included appropriately.

In addition to these publications the Secretary's annual report took on a decidedly literary flavor. Before his arrival in office the annual report to Congress had been a mere record of receipts and expenditures, but the Committee on Organization recommended that "as an additional means of diffusing knowledge, your committee suggests the publication of a series of reports, to be published annually or oftener, containing a concise record of progress in the different branches of knowledge, compiled from the journals of all languages, and the transactions of the scientific and learned societies throughout the world." [15]

For the first thirty years this report was limited to 450 pages. The number of copies printed varied considerably, but the average edition was 15,500 copies. It is difficult to epitomize the contents of these volumes, but they are familiar to all students of research. Articles and papers published abroad were translated and made accessible to many more readers. Noteworthy among the translations were Müller's reports on "Recent Progress in Physics," in which the famous Freiburg professor treated such subjects as galvanism and static electricity during the years 1853–1858. The advances made by such eminent workers as Joule, Helmholtz, Thomson (Kelvin), Maxwell, and Hertz were promptly recorded.

An interesting sidelight on Henry's character was revealed when the suggestion was advanced that all the Smithsonian original works should be protected by copyright. He decided instantly that all the knowledge the Institution should be instrumental in offering to the world should be free to all who were capable of using it. There are no copyrights.

He was not content that the mere publication of printed knowledge would ensure its satisfactory diffusion. Not the least noteworthy

[14] Fulton and Thomson. *Benjamin Silliman*, p. 124.
[15] *Smithsonian Report for 1846*, p. 23.

aspect of his work in this direction was the very effective international exchange system of publications he established. He cannot be credited with having originated the idea. Some tentative steps had been taken as early as 1694, when the French royal library had exchanged duplicate books for new books printed in foreign countries. Later, the American Philosophical Society and the American Academy of Arts and Sciences had instituted limited exchanges. A good deal of prominence had been gained by the efforts of Alexandre Valtemare, a French surgeon and professional ventriloquist, who succeeded in starting exchanges between the United States and France. All these efforts had been on a small scale. Henry's plan was to give the system the widest expansion.

He began by arranging for the exchange of the Smithsonian publications for those of other learned societies, but once the distribution of the first memoir had been accomplished it became evident that the chain of exchange could be enlarged. The privilege of sending and receiving parcels of books through the Institution was extended to other American institutions and, in some cases, to individuals. No charges were made for sending moderately sized parcels from Washington. Henry was very proud of his system of exchanges once it had been set in operation. Speaking on the subject he said:

> The worth and importance of the Institution are not to be estimated by what it accumulates within the walls of its building, but by what it sends forth to the world. Its great mission is to facilitate the use of all the implements of research, and to diffuse the knowledge which this use may develop. The Smithsonian publications are sent to some institutions abroad, and to the great majority of those at home, without any return except, in some cases, that of co-operation in meteorological and other observations.[16]

The system had been in operation for twenty years and had reached its maximum volume when, in 1870, Henry was invited to give testimony in London before a Royal Commission on scientific education. In answer to a question on the purpose of the exchanges put to him by one of Commissioners, he said:

[16] *Smithsonian Report for 1852*, p. 20.

FOUNDING THE SMITHSONIAN

The great object is to facilitate in every way the promotion of science, and especially the fostering of original research, and enlarging the bounds of human thought. It is a matter of surprise that the idea is not more generally understood by statesmen and legislators, that modern civilization depends upon science, including the knowledge of the forces of nature, and the modes in which they become the agents of man.[17]

An appreciation of the work came from an unexpected quarter. Several of the steamship lines esteemed the exchange system so highly that they readily acquiesced to Henry's request that they transport the packages without charge for freight.

As the system proved its efficiency several foreign governments saw in it the appropriate channel for exchanging official documents with the United States government. Arrangements were speedily concluded and the flow of packages was started which amounted to 99,000 pounds of literature in the thirtieth year of the enterprise. This international exchange of government documents was conducted through the agency of the Smithsonian until the year 1881.

The first annual report of the Secretary to the Regents mentions the need for the Institution to take the lead in establishing an extensive system of meteorological observations, particularly with reference to the origin and phenomena of American storms. Here Henry was continuing an early interest which he never outgrew. His efforts to promote the accumulation of weather data were to obtain for him the title of "Father of the Weather Bureau."

Although he could not have been aware of it, he was entering upon a field which embraced some of the most involved and intangible phenomena in the whole field of science. This could not be apparent in Henry's day, since no approach could be made to the numerous problems until the fundamental facts had been observed, assembled, and investigated. The task which he undertook was to organize a system whereby the observations could be made so that the assemblage of facts would permit the variables to be noted and, if possible, be defined. This work has not yet been completed, but without the foundation laid by Henry, no progress could have been made. It is to his credit that a start was finally made in this country to organize

[17] Rhees, W. J. *Journals of the Board of Regents,* etc. Washington, 1879, p. 783.

meteorological observations on a national scale, and it was through his influence that it functioned smoothly and efficiently from its inception.

There were no means in existence of foretelling the approach of a storm, except such as could be guessed by isolated observers on the spot, who had no means of communicating their forecasts to anyone but nearby residents. Since there were no known laws relating to revolving storms, all forecasts were the result of guesswork.

In urging upon the Regents the wisdom of undertaking an extensive organization of weather observers, Henry wisely obtained the weighty opinions of the most prominent meteorological authorities of the day, men like James P. Espey and Elias Loomis, which supported his own.

A close friendship had sprung up between Espey, "the Storm King," and Henry. At the time of his death, the former left his friend a precious telescope. Some years later, Henry found himself the dinner companion of Espey's niece. They talked about old associations and Henry delivered the opinion that "There is no question in my mind but that Professor Espey should be regarded as the father of the present signal service of the United States, his theory of storms having led the way to the establishment of the service and its present success."

This was an example of the generous tributes Henry paid to other men's achievements, deprecating his own contribution and magnifying what another had done. But his judgement need not be taken too seriously. It was Henry, and he alone, who perceived the practical way of collecting the data, and of assembling and correlating it. Until this was done, the material could have nothing but a local significance.

The Regents were persuaded to appropriate the sum of $1,000, to be spent upon instruments for distribution among volunteer observers.

In 1848 Arnold Guyot came to this country from Switzerland. Henry met him in Philadelphia at a meeting of the American Association for the Advancement of Science and, perceiving his worth, consulted him upon the project. It was desired to secure simultaneous observations upon barometric pressure, air temperature, humidity, wind velocity, cloud formations, and precipitation from as many observation posts as possible. Where measurements were to be made it was essential that the instruments adopted should be uniform in design and in accordance with the latest ideas for ensuring the greatest accuracy. Guyot was persuaded to undertake the task of choosing and in some cases of making the necessary instruments. His wisdom was largely responsible for making the small sum appropriated go a very long way.

Until the electric telegraph had been invented there was no means of obtaining reports from several distant observation posts in time to foretell the arrival of a storm until it was ready to break overhead. Now that Morse's telegraph was available, Henry proposed to organize the corps of voluntary workers having access to the telegraph into a reporting network. From their telegraphed reports of observations a daily weather map might be plotted at the Institution.

After the plan had been put into execution on a limited scale and the promise of success began to be realized, Henry began to seek the means of extending the service. In 1849 he attended the meeting of the A.A.A.S. in Boston, and a happy inspiration prompted him to extend his journey to Toronto. Here he was warmly welcomed by Captain J. Henry Lefroy, who was in charge of the Observatory. Lefroy promptly promised him the support of all the British observers in Canada, Bermuda, the West Indies, and Central America, who had been organized as the outcome of the British Association's suggestion in 1838, and which the American Philosophical Society, on Henry's suggestion, had asked the United States government to join.

Thus, at this very early stage of the scheme, Henry's cooperative network was spread over the greater portion of North America. By close correlation of reports received from all these widely scattered observers and the comments of meteorological experts like Espey, Loomis, and Coffin, it became possible to assemble enough information to perceive general laws relating to the weather and to plot the course of revolving storms. Henry immediately saw that this organization would serve the practical purposes of agriculture and shipping while adding to the sum of human knowledge.

The information contained in the daily reports which poured over the telegraph was plotted on a large map of the United States by means of adjustable symbols. This was the original daily weather map. "The first published weather forecasts were based upon telegraphic simultaneous observations inaugurated by Professor Henry of the Smithsonian Institution in 1849, . . . where they were daily exhibited on a large wall map until the year 1861. These reports were frequently used by Professor Henry to predict or show the possibility of predicting storms and weather; a matter that he had frequently urged on the attention of Congress." [18]

An interesting picture of Henry as a practical weather forecaster has been drawn by John Wise, the balloonist.

> "Old Probabilities"—I mean the original one, Joseph Henry, gave me a lesson in this weather prediction sixteen years ago. With his weather map before him and a telegram from St. Louis describing a storm there and then in action, he began to trace it by sticking pins into the map. As we had been discussing the nature of the storms and the upper trade winds the evening before, I remarked, "You ought to be able to tell when that storm will reach the meridian of Washington," and he did predict it truthfully.[19]

Wise had made a balloon ascent in 1859 to make meteorological observations at Henry's suggestion, and he had named his balloon the "Smithsonian."

In 1851 the corps of observers was invited by Henry to supply information on the flowering and fruiting of certain plants, the first appearance of birds, animals and fishes, and the opening and closing of navigation. The observations made over a period of six years (1854–1859) were tabulated and published under Smithsonian auspices in two large volumes. Henry himself interpreted their meaning in a series of articles published in the annual reports of the Patent Office, beginning in 1855, under the general title of "Meteorology in its Connection with Agriculture." [20]

Today, when books dealing with specialized knowledge are plentiful and inexpensive, we tend to overlook the valuable papers with which the Patent Office and the Census Bureau used to embellish their acres of statistics when technical books and journals were fewer in number. Henry's efforts to popularize the science of meteorology

---

[18] Weber, G. A. *The Weather Bureau*, 1922, p. 2.
[19] Wise, J. *Through the Air*, 1873, p. 416.
[20] *Agricultural Report of the Commissioner of Patents*. Washington. Part 1. General Considerations, 1855. Part 2. General Atmospheric Conditions, 1856. Part 3. Terrestrial Physics and Temperature, 1857. Part 4. Atmospheric Vapor and Currents, 1858. Part 5. Atmospheric Electricity, 1859.

may have been buried from the eyes of all but the more diligent searchers, but they were not altogether lost upon his contemporaries, and deserve more attention from modern readers. These essays constitute the first scientific treatment of the physics of the air in this country, and they contained material which must have been of great interest to the students of meteorology at the time. In the opening remarks Henry wrote: "Independently of the practical value of a knowledge of the principles on which the art of agriculture depends, the mind of the farmer should be cultivated as well as his fields, and after the study of God's moral revelation, what is better fitted to improve the intellect than the investigation of the mode by which he produces the changes in the material universe?" [21]

The third article in the series, "On Terrestrial Physics and Temperature," presented his readers with the enlarged view of his ideas upon the constitution of matter and the correlation of forces which should have been given to the world through some other vehicle with a wider distribution, together with a clear statement that it was an elaboration of views expressed at an earlier date. This might have enabled Henry to regain the position he had lost (or, more truthfully had never gained) as a pioneer in the movement to correlate certain natural forces.

The meteorological service which Henry had started grew until it embraced six hundred observers, whose services were called upon on occasion for additional information upon natural history, topography, archaeology, etc., relating to their immediate neighborhood. It was Henry's intention to supplement the information they furnished by the collection and examination of records on meteorology assembled by various societies for the purpose of establishing general laws. The outbreak of the Civil War put an end to this project, and when it was revived, the work had been transferred to a government sponsored agency.

All the time that he was writing articles, devising methods, providing instruments, preparing mathematical tables, and enlisting the services of others to extend the studies, he was conscious that the country had need of a centralized weather bureau with a larger income than the Smithsonian could provide. He began to agitate for the formation of a national weather bureau in 1865, but he had to wait for five years before his wishes were partially realized when a

[21] *Agricultural Report*, 1855, p. 357; *S.W.* Vol. II, p. 11.

separate department was created under the Signal Office of the War Department.

The work on meteorology we have described, beginning in Albany, does not cover all Henry's activity toward the advancement of this science. He furnished the article on the subject (one on magnetism, too) for the *American Cyclopaedia* in 1861, and he deserves to be better known for having instituted the now universally adopted system of storm warnings. He also planned the atlas of rainfall and temperature for the Census Bureau.

The subjects in which Henry displayed an interest are bewildering in their variety, but he was wise enough not to attempt to perform the expert's work. When a specialist's knowledge was necessary he took care to seek out a recognized authority and, having enlisted his service, the Secretary allowed the specialist to conduct the work, limiting his own participation to external support and supervision.

Professor Spencer Baird was granted aid to make an extensive study of birds; Agassiz to study fishes; and again Baird was encouraged to undertake an investigation into an alleged decrease of food fishes in the waters around the coasts.

These and the numerous other studies belong to the story of the Smithsonian rather than to the story of Joseph Henry, but he entered more directly and influentially into many other activities of the Institution which seemed just as remote from his particular interests.

At the time when he assumed duties as Secretary, more than a fourth of the globe was practically unknown territory, represented on maps as bare outlines enclosing conspicuous blanks. Henry began to take a keen interest in the efforts to explore the comparatively unknown territories. The funds of the Institution did not permit of the outfitting of any elaborate expeditions, but when small sums were available they were devoted readily to pay partially the cost of an expedition that would bring back knowledge of some little known part of the world. The most conspicuous results were obtained from the more readily accessible territories of the United States which had not been thoroughly explored.

Shortly after the Institution had begun operations, the federal government inaugurated a series of surveys having as their purpose the delimitation of state boundaries and the survey of projected railroad routes. These were followed by topographical, biological, and geological expeditions in the interest of commerce and fisheries. As the outcome of these explorations, a fuller understanding of the country's natural resources was obtained at government expense.

Henry and his staff played an influential part in these undertakings. The government's participation in scientific research, even of a practical nature, was much less extensive than it has since become. If public interest has shown a tendency to expand we have to thank the wise and efficient work of Henry and his associates in the initial stages. In some cases the Institution took the initiative, and the government was only induced to take part when the practical bearings of the expeditions had been demonstrated.

This is a little known phase of the Institution's work, since it was never Henry's method (nor that of his successors) to seek the credit for the management of explorations. Satisfaction was derived from aiding others to labor in the cause of science. The first publication of the Institution, we have pointed out, was the outcome of an exploration of ancient monuments; and in the next few years there were reports, usually from surveys and expeditions which promoted the study of ethnology, adding materially to the sum of knowledge concerning the American continent.

The annual reports of the Smithsonian Institution for the twenty years between 1851 and 1871 contain a comprehensive history of the work undertaken by the government in the exploration of the new regions in the West. They constitute, in fact, the only systematic record in existence of the explorations covering this period. The task of preparing instructions for the explorers, of arranging their scientific equipment, and inspiring the workers with enthusiasm was shared with the Assistant Secretary, Spencer F. Baird, but Henry supervised it all.

Henry and Baird prepared the instructions for the scientific observations to be made by Captain Hall's famous Arctic expedition. Advice was freely given to all such expeditions fostered by the government, and when expeditions were fitted out by private enterprise, Institution officials sometimes accompanied them to conduct the scientific aspect of the exploration. A notable example of this practice was the Western Union Telegraph Company's exploration of Alaska and Eastern Siberia for the purpose of building telegraph lines overland to Europe. The printed reports of the expedition were its principal outcome, for the project of the telegraph was abandoned after the trans-Atlantic cable had been successfully laid.

When officials were not accompanying the expedition, its geological, botanical, and ethnological experts were instructed at the Institution. The magnificent series of Pacific Railroad reports owed much to the cooperation of the staff at the Institution.

One survey had an unusually wide practical value to the economy of the nation. The annual inundations of the Mississippi Valley were often the cause of ruinous destruction to the cotton and sugar crops, while the bar at the mouth of the river was a serious impediment to navigation. The surveys of the Mississippi and Ohio Rivers by Charles Ellert were under the general supervision of Joseph Henry. The reports were instrumental in adding to the knowledge of the physical geography of the territory, and served as an invaluable guide in determining the steps to be taken to improve the navigation of the rivers and to prevent inundations.

In the summer of 1867, a modest exploration of the Colorado canyons was fostered by the Smithsonian. Major John W. Powell and a group of associates collected around Henry and Baird to receive instruction. This was most fortunate because, although the expedition had been designed to be modest and amateurish, it proved in the event to be the boldest in design, the most perilous in execution, and among the most fruitful of all expeditions undertaken in this country.

It began as a geographical survey, but the members were so admirably instructed that it embraced geology and biology. Arrangements had to be made to extend the scope of the expedition as it advanced, and Henry was largely instrumental in obtaining government funds which enabled the explorers to pursue their work. His aid to this expedition was commemorated in an enduring monument in the name given by the explorers to the Henry Mountains in Utah.

Springing from this expedition was a well-conducted research into the study of the aboriginal inhabitants. An extensive collection of objects was made illustrating the arts, languages, and institutions of the Indians for preservation in the Smithsonian. In 1876 a series of reports on the Indians entitled "Contributions to North American Ethnology" was projected by Henry, and the first two volumes were issued in the following year.

By this time the survey had been transferred to the Department of the Interior, and it was judged expedient for the Smithsonian Institution to withdraw from active participation. All its collections of material relating to Indian civilization made during the preceding thirty years were transferred to the new organization which became, by Act of Congress in 1879, the Bureau of American Ethnology. Among the collections handed over were 670 Indian vocabularies, indicative of the intensive manner in which this phase of the Institution's activity had been pressed.

Not all the expeditions in which Henry and his staff were called upon to advise related to the United States. Much valuable work was accomplished in the Hudson Bay Territory, while explorations of Yucatan, British Honduras, the Commander Islands, Korea, China, and Tibet were conducted under the Institution's auspices.

The extent of the Institution's collaboration will not be found complete in its own publications. Acknowledgement of the aid rendered, too frequently, must be disinterred from the reports of government bureaus and of societies.[22]

Henry did not limit his labors to the aid given to the scientists who accompanied these expeditions. The sciences of archaeology and ethnology were new to the mass of the American people, who had been too fully occupied with the problems of the present to devote much time to the study of the past. Yet they were being asked to pay from public funds the costs of many surveys which, although they had practical purposes, were concerned with the collection of the information and material upon which the new sciences were being nourished. In order that the general public might be better instructed upon the value of these archaeological and ethnological studies, Henry strove to broaden the interest in the subjects by preparing and distributing through the Institution's facilities popular circulars designed to appeal to the widest circle of readers.

In his attitude toward archaeology and anthropology, Henry showed a liberalism that is noteworthy among scientists. The origin of the races of man, of man himself, of life and of species, the hypotheses of the anthropologists and archaeologists were matters which the laboratory scientists had too often regarded as incredulous, as not belonging to the true domain of science. Henry perceived the

[22] Brief accounts of the principal expeditions assisted and the ethnological results derived will be found in *Smithsonian Institution: 1846–1896*.

wisdom of applying the scientific method in these regions, and he showed his readiness to encourage anyone who embraced their investigation in the scientific spirit.

During the period he had worked in experimental research Henry had suffered doubly through the inadequacy of a knowledge of scientific literature. The greatest smart may have been to his pride, through the neglect by Europeans of his valuable contributions to science. He might not have found it difficult to offer excuses for those men who had failed to give him the recognition he felt was his due. After all, how many Europeans had access to the two journals in which his work had been published?

The other source of suffering had sprung from his difficulties in not having before him a full survey of the subject which occupied his mind at the moment. He, like all his compatriots, was tormented with the thought that the very incomplete libraries to which they had access did not furnish all the material they might require to spare themselves from making investigations which might have been performed very thoroughly by other workers in the same field, but whose labors might remain unknown. What was urgently required was a comprehensive survey of the literature of science which would inform workers all over the world what had been accomplished in their various fields.

The literature had been fairly well catalogued down to the end of the eighteenth century, but the great volume of material which had since been published, especially in periodicals and annual reports, was awaiting an index. Henry conceived the idea of making a catalogue of scientific papers. He broached the matter in his report for the year 1851.

> One of the most important means of facilitating the use of libraries is well-digested indexes of subjects, not merely referring to volumes or books, but to memoirs, papers, and parts of scientific transactions and systematic works. . . . Everyone who is desirous of enlarging the bounds of human knowledge, should in justice to himself as well as to the public, be acquainted with what has been done in the same line, and this he will only be able to accomplish by means of indexes of the kind above mentioned.

His plan of undertaking the indexing of scientific literature was magnificent, but when the proposal was examined in detail and work could be foreseen extending over many years for the united labors of a large group of especially trained bibliographers, the question of finding funds for the ambitious enterprise was seen to be unanswerable. Henry reluctantly abandoned the project when he realized that one half the Institution's income was appropriated for the support of what he regarded as undesirable departments which Congress had inflicted upon it, and the remainder was wholly insufficient to provide funds for the contemplated undertaking.

But the idea was much too good to abandon altogether. In 1854 Henry could see his way clear to undertake the task of indexing American scientific literature if some other organization could be induced to cooperate so far as the more extensive British and continental European literature was concerned. Accordingly he proposed to the British Association that it was a task worthy of their consideration. The British Association welcomed the proposal and appointed a committee to consider the best means of executing it.

By chance or design the three members of the committee were Fellows of the Royal Society. It was their opinion that this latter body was best qualified to undertake responsibility for the work, and they succeeded in arousing its members to become interested in the project. However, the task was regarded to be beyond the means of the wealthiest society, but so clear was the need for such a valuable bibliography, that the Fellows prevailed upon the British government to make them a substantial grant which would permit the discharge of a duty to contemporary science.

Thus the way was prepared for the compilation of the monumental "Catalogue of Scientific Papers," the first volume of which appeared in 1863, ten years after Henry had proposed it to the British Association. The first volume bore testimony to the origin of the idea, for it announced: "The present undertaking may be said to have originated in a communication from Dr. Joseph Henry, Secretary of the Smithsonian Institution." Thus, a generation of scientific research workers were notified of the identity of the man to whom they were indebted for one of their most valuable tools.

This effort did not exhaust Henry's interest in the provision of indexes to scientific literature. He knew better than any other man in America how gigantic was the enterprise and, knowing the prevailing need, he sought to furnish a temporary alternative for the specialist. He was instrumental in furnishing through the Institution's publications so many special bibliographies that they are too numerous to speak of individually.

It should not be forgotten that Henry furnished the initial impulse to that bibliography of science which has grown in importance with every succeeding year, for it is now recognized as the key to scientific research. The quick and certain finding of what has already been discovered is almost as important in some branches of science as a new discovery; it saves a vast amount of duplication, and it reduces the liability of the research worker to repeat errors.[23]

An activity of a totally different nature centered around the foundation of a national herbarium. The close friendship existing between Henry and John Torrey, who was a botanist before all else, must have aroused in the former a desire to know more about herbs and plants than he had acquired while a student of "Physik," but he had never displayed a marked inclination toward that subject. However, the botanists under Torrey and Gray had carried their science to a stage which commanded international respect. It had been Henry's hope and intention to have Torrey with him at the Smithsonian Institution, but as the project of having a staff of professors never materialized, Torrey remained the consultant on the subject of botany. The two old friends had been working toward the establishment of a national herbarium for years, and their hopes came near realization in 1860 when Henry laid claim to the collections of plants made by the various exploring expeditions conducted under government auspices. In 1870 all the collections from the expeditions and several from private sources were turned over to him for classification in one great collection. Today this great collection, housed in the National Museum, is approaching a total of two hundred thousand specimens.

While carrying on the multifarious tasks of organizing and administering the Institution which was engaging in activities of a nature entirely unlike anything he had hitherto attempted, Henry was called upon to exhibit qualities which did not appear to be compatible with the character of a pure scientist. Not the least astonishing revelation

---

[23] The neglect of research workers to take advantage of these invaluable tools causes the loss to American industry of several million dollars a year. See the author's "Neglected Aspect of Research." *Jour. of The Franklin Institute.* Vol. 241, March 1946, p. 187.

of his unsuspected qualities were his operations in the role of financier.

The student of history or economics might have recalled with some trepidation the lamentable failure of the great mathematician Laplace who, as Napoleon's Minister of the Interior, had attempted to solve the problem of the French national debt by means of the infinitesimal calculus. Such a dismal example might have provided justification for Henry's friends in regarding with misgiving his inexperience in the handling of a financial trust. It was no minor triumph when he demonstrated his ability to control the expenditure of the funds in his charge with a skill that put to shame those who may have argued that a knowledge of the rigid rules of science would not serve where the flexible principles of finance were concerned.

In finance, Henry never experienced any difficulty. If he had any financial policy it was a compound of his inheritance from a line of provident ancestors and of experience gained in the humble home of his widowed mother. His financial theory was reduced to the two simple rules: pay as you go, and spend less than your income.

The sum bequeathed by Smithson amounted to $515,169. At the time of Henry's death the funds of the Institution in the hands of the Treasury amounted to $686,000, while the real estate, library, equipment, and stock of publications were valued at an additional $782,000. Truly a flattering picture of an amateur financier's ability!

But it is not to be supposed that Henry's life in Washington was an unbroken triumph, without its rough spots and an occasional unpleasant jolt. His benign nature would have carried him through a private life without any untoward sign of friction, but fate had made him a public servant and the defender of a policy. Men who occupy public office in the United States rarely escape from energetically delivered attacks by critics of the policy they have to administer. Henry was no exception.

Too often in his life, the casual student will see him accepting with apparent meekness the buffets of misfortune. None but a careless student would attribute this attitude to submissiveness. It was pride. Henry was submissive before nature, but not before man. Once his mission was clear he could and did struggle with dogged obstinacy to gain his ends. This became abundantly clear as he was observed fighting for the principles of his plan of organization for the Smithsonian Institution. To the attainment of that plan he brought an unwavering unselfishness, a pertinacity that conquered opposition, and a wisdom that disarmed opponents. But he had to wage a long and bitter struggle and he did not live to see the complete fulfillment of his vision. His battles deserve a chapter to themselves.

THE TELEGRAPH CONTROVERSY

As we have seen from his Christmas letter to President Nott, Henry was not without foreboding when he took up the task of organizing the Smithsonian Institution. He realized that he had been called upon to execute a policy laid down by Congress that was fundamentally opposed to his own ideas. One might wonder why he accepted a post with so many obstacles.

There can be little doubt that he assumed the task with a feeling that it was a duty to remove the perversions of Smithson's wishes, and that if he failed he would resign his office rather than attempt to execute a policy which he believed to be wrong. The strength on which he relied lay in his power of wearing down opposition by sheer weight of patience and reasonableness. Unable to gain a favorable decision at the beginning, he was willing to compromise while gaining the confidence of the Regents.

It was his hope that he would ultimately succeed in persuading them to furnish the support he needed in divesting the Institution of those outcrops that Congress had permitted to grow upon Smithson's idea. He was prepared to be strong in defense of his own policy, and wise enough to proceed warily in demolishing the policy Congress had imposed upon him. Above all, he was unremitting in his watchfulness against any attempt to encroach upon the Institution's funds.

This brought him into conflict with some redoubtable opponents who sought to attach the funds to their own interests, but the purity and the simplicity of his character and motives foiled, as no other protection could have done, the wiles of the politicians who would have used the funds for other ends.

One of the most direct attempts to plunder the Smithsonian funds was led by Stephen A. Douglas, then at the height of his influence. Not yet forty years of age, the "Little Giant of Illinois" had just defeated General Case for the presidential nomination but, by throwing the weight of his influence to Franklin Pierce, he appeared to have secured his claim to nomination as the candidate for the 1856 election.

Douglas was seeking the agrarian vote. In order to gain it he was not above inventing a political stratagem. He arranged for an Agricultural Convention to be held in the lecture hall of the Smithsonian in the year 1852, at which time it was proposed to petition Congress to form an Agricultural Bureau. Such a proposal would have been entirely praiseworthy had not Douglas conceived the idea that the necessary funds to establish the bureau should be furnished from the Smithsonian bequest. A raid upon an apparently defenseless institution's treasury appealed to predatory political instincts; it was a shrewd move to provide the farmers with an advisory department without having to dip into the taxpayers' pockets.

Douglas was a formidable platform opponent, a debater of high reputation, as Abraham Lincoln knew, but Henry was not afraid to challenge him on his own ground.

When the day came, Douglas advocated a raid upon the Institution's funds in a speech typical of the politician eager to play on the selfish wishes of his audience. Civilized man, he contended, was dependent upon agriculture, without which a country would descend into barbarism. If Smithson had left a sum of money for the promotion of American civilization how better could this object be served than by dispensing the money for the benefit of those at the foundation of that civilization?

Henry occupied an inconspicuous seat at the back of the hall, where he listened attentively to the proposed raid upon the funds of which he was the guardian. The applause which greeted Douglas's proposal demonstrated how popular it was with his hearers. The fact that he had to face an unsympathetic audience could not deter Henry from the discharge of his duty. Much to the surprise of the promoters of the meeting, who had anticipated no disapproval, Henry rose to offer opposition to the "Little Giant."

In a calm and dignified manner he introduced himself as the

Secretary of the Smithsonian Institution, official protector of the funds they proposed to seize. Addressing his words beyond the scheme's advocate to the farmers themselves, he expressed confidence they would not consciously strive to turn to their personal advantage benefits intended for all men. Turning from the moral to the legal aspect of their claim, he denounced the contemplated breach of trust. Here, the asperity of his remarks wounded Douglas's pride. Henry then concluded his speech with a reminder that new truths and their subsequent applications in the useful arts, the promotion of which was the Institution's function, had been the means by which the farmers had been raised from the position of drudges to their present stage of comfort; to attempt to reduce the Institution's power to uncover new truths and to publicize them, would be a very short-sighted policy.

After the debate, which was not without further asperities, the two principal protagonists met outside the hall, where they exchanged apologies for the harsh words they had employed, and parted good friends. Douglas carried away from this encounter a high regard for his opponent. He knew that he had met his match. Nothing more was heard of the proposal to start a farm bureau at the expense of the Smithsonian.

In 1854 Douglas was elected a Regent of the Institution, and retained the office until the time of his death in 1861. During this period he came to acquire a better appreciation of the Institution's work and learned to understand both it and the character of its secretary. Henry, too, got a better understanding of the fiery senator.

Unconsciously, Douglas did the Smithsonian a service. His failure served as a warning to less gifted performers that it was unsafe to contemplate raiding Smithsonian funds.

Not all of those who opposed Henry's wishes in regard to the nature of the Smithsonian's functions were animated by questionable motives. Among those whose opinions deserve to be treated with great respect were those men who sincerely believed that the establishment of a national library comparable to the British Museum in London or the Bibliothèque Nationale in Paris was the surest way of carrying out Smithson's wishes.

The great mass of the American people were uninterested. The nation was still looking to Europe for leaders in science, and had produced few great scientists. On the other hand, literature in this country was experiencing a brilliant revival, and for the time being the claims of the writers far outweighed those of the scientists for a place in public esteem.

There was no real struggle between science and literature, but in the public mind the claims of the two had to be settled. In such a situation literature begins with a head start. It can, and does, appeal directly to the public, whereas science rarely touches many people directly. The multitude read books, and the period between 1850 and 1855 is one of the richest in the whole history of American literature. It witnessed the appearance of *Representative Men, The Scarlet Letter, The House of the Seven Gables, Moby Dick, Walden,* and *Leaves of Grass*. Although few of these may have made immediate appeal to the mass of the nation, there can be small wonder that the reading public would fail to be impressed by the claims of literature to recognition.

American science had no such achievements to set alongside these masterpieces. The future of the American intellectual effort must have appeared to rest with the literary rather than with the scientific workers. This made doubly difficult Henry's task of convincing the library supporters of their error in striving to acquire the major part of the Smithsonian funds. Against this class of opponents he employed a patient sagacity.

He was not opposed to the idea of a worthy national library, but he was strongly opposed to having it imposed upon the Institution as an obligatory function. He based his opposition upon two arguments. A large library situated even in the national capital could have only a local advantage, whereas Smithson's desire was to serve all men. Furthermore, the entire income from the bequest was insufficient to support a library of the size contemplated.

The Salem lawyer Senator Rufus Choate, an ardent book collector whose house was filled with books from cellar to garret, had been the staunchest advocate of "a noble library, one which for variety, extent, and wealth shall be confessed to be equal to any now in the world." He had wished to have inserted in the founding act a clause ensuring that not less than $20,000 be spent each year on the purchase of books. This was anathema to Henry, who foresaw that a similar annual sum would be required for cataloguing, binding, and repairs, thus exhausting the Institution's income.

Congress did not accept Choate's proposal, but it did permit insertion of a clause that provided for the payment of "not exceeding an average of $25,000 annually" for the gradual formation of a library. This permissive clause was to lead to trouble.

Henry persuaded the Regents that the appropriation of this amount to be spent on books would cripple the Institution's activities. He proposed that not more than one half the amount be devoted to book-buying. Even this was in the nature of a compromise, for he never ceased to urge that the Institution should be freed from the incubus of a large library. When he proposed a further reduction in the book fund, opposition flared. Choate and Representative Meacham of Vermont, both Regents, and C. C. Jewett, the Institution's librarian, all took part in leading the opposition to Henry's proposal.

Jewett was a librarian of the first rank, and he desired to take the fullest advantage of the law's provision to make his department strong. He fully recognized that the expenditure of $25,000 a year on books would consume the major portion of the income, and he gave an ungracious consent to the arrangement whereby half that sum should be apportioned to the support of the library and the museum, while the remainder of the income was devoted to the "active operations" which Henry regarded as the only legitimate functions of the Institution.

Jewett grew restive under this arrangement, he was impatient to become head of a noble library such as Choate had projected. He was not averse to the library tail wagging the Smithsonian dog, since this would exalt his own position above that of the Secretary. Unfortunately, he could not repress his impatience. He demanded that the maximum sum mentioned in the Act of Congress be given to the library. If the demand were refused, he unwisely threatened to shake the Institution to its foundations.

Chafing under what he thought was the inferior position into which he was being thrust by Henry's action, Jewett rashly went over the Secretary's head in addressing to the Regents a demand that the library receive all the money to which he felt it was entitled. In this demand he overstepped discretion in criticizing Henry's qualifications to hold the position of Secretary, and ridiculed the annual reports he submitted to the Regents. To have suffered this act of insubordination to pass without prompt action would have put an end to the Secretary's authority over his staff.

Not without reluctance, for he had a high respect for Jewett's professional abilities, Henry dismissed the offender. This precipitated a battle.

Jewett retaliated by challenging Henry's authority to dismiss him. A special committee of the Regents was appointed to investigate the dispute between the Secretary and the Librarian. Their report vindicated Henry's action.

This report angered Choate. No longer a member of the Senate, he once more projected his presence into its chamber through the instrumentality of a letter of resignation from the Board of Regents. Although the reason he offered for resigning was his residence in Boston, which rendered periodical visits to Washington inconvenient, he went out of his way to express disagreement with the Regents' attitude toward the library, and disapproval of their delegation of executive authority to the Secretary. His action met with some support, and the conduct of the Regents was submitted to investigation by the Judiciary Committee of the Senate, the members of which reported that they could find no grounds for complaint.

But the advocates of the large library refused to admit defeat. Representative Meacham, also a member of the Board of Regents, requested the appointment of a House investigating committee. As one reads his speech [1] it becomes apparent that Henry had failed to convince all the Regents that his plan was sound. The Vermont representative was a practical gentleman who wanted members of Congress to have something substantial to show their visitors in Washington, and a library of imposing proportions would be visible evidence that wisdom prevailed in the capital.

Nor was Meacham an admirer of Henry, whom he described as being "smitten by chronic megalomania on a single subject." He was not in the least impressed by Henry's grander conception of Smithson's purpose. He found satisfaction in the New York *Tribune*'s inept description of the Institution as a sort of "lying-in hospital for literary valetudinarians."

The members of the investigating committee appointed by the House did not regard their duties very seriously, and gave satisfaction to no one. The "majority" report was signed by a single member, the chairman, who condemned the Regents. The "minority" report was signed by two members who exonerated the Regents and Henry in

---

[1] *Congressional Proceedings, House of Representatives,* March 3, 1855.

their actions. The other two members were so indifferent toward the issues involved that they expressed no opinion.

These proceedings must have been highly distasteful to Henry. Bad enough that a colleague chosen for his undoubted ability should have been unable to work in harmony with him and that the discord should have provoked Congressional debate. But this was not the whole story. He had been deriving a quiet satisfaction from winning over a majority of the Regents to his plan and to the evasion of the unwise provisions of the Act of Congress in founding the Institution. Now these men were exposed to the humiliation of having their confidence in him exposed to public criticism, for the dispute did not end with the reports of the investigating committees.

Choate's disclosure that not all was harmonious at the Smithsonian provoked an acrimonious discussion in some newspapers which degenerated into a senseless wrangle over the relative claims of science and literature. During this ordeal, the towers and shaded paths of Princeton must have beckoned to Henry with almost irresistible invitation to return. Had the course he was following not received at least partial vindication nothing could have induced him to retain his position. But when the majority of the Regents stood loyally by him and gave unequivocal support to his work, he could not think of surrendering his trust. He must execute his plan to the fullest extent within his power, now that the preliminaries had been cleared away. But there were still obstacles to be surmounted.

Then, when the clouds were beginning to disperse, the cruellest blow of all was dealt him.

We have seen how in 1837 Henry had withheld any announcement against Morse. It will be recalled how he had been urged by friends to declare that, by the publication of a paper containing an account of his experiments in transmitting electrical impulses over a long line, the fundamental principles of the electric telegraph had been thrown into public domain. He was advised to protest when Congress aided Morse, and to encourage it to appropriate funds to reward the inventor of the best system embodying those principles. His ingrained habit of refraining from doing anything that could be postponed had caused Henry to neglect writing the letter to dissuade Congress from subsidizing Morse. He then met Morse and became convinced that,

THE TELEGRAPH CONTROVERSY

as he had developed a sound method for building a telegraph, he was worthy of encouragement.

Once Professor Gale [2] had set him on the right path, Morse had sought Henry's counsel, approaching him with the deference of a grateful pupil asking a favor of his instructor. Upon the first of his visits to Princeton, Morse had been informed of the "combined circuits" which made possible the transmission of signals over long distances. Henry had also acquainted Wheatstone with this essential feature of the telegraph while on his visit to London. Henry always regarded the relay as his own invention.

"One morning, he [Henry] came into my laboratory at Cambridge," wrote Professor Trowbridge of Harvard, "and after I had shown him several pieces of scientific apparatus, he stood before an electromagnet which was working a relay, and looked long at the magnet, and then at the battery which was coupled for quantity, and remarked in a quiet tone, as if to himself, 'If I had patented that arrangement of magnet and battery, I should have reaped great pecuniary reward for my discovery of the practical method of telegraphy.'" [3]

The benefit Morse derived from his acquaintance with Henry was again made apparent in 1842, when by his own confession the inventor was at the bleakest moment of his career and was struggling, almost without hope, to secure aid from Congress which would enable him to construct a telegraph line. He was advised to obtain letters from scientists approving his plan. On January 11 he wrote to Henry begging him to write a letter urging Congress to give him aid. Henry answered on February 2 with a letter saying: "Science is now fully ripe for this application, and I have not the least doubt, if proper means be afforded, of the perfect success of the invention."

This was the brightest ray of hope that penetrated the somber surroundings of the inventor. The letter was a tonic, and he made full use of it, as Henry had intended, in employing it to canvas support among members of Congress. He wrote dozens of letters quoting Henry as "a high authority" who approved of his scheme.

This was no misrepresentation. Henry missed no opportunity to

---

[2] Dr. Leonard D. Gale, professor of chemistry at New York University, where Morse was a professor of fine arts.

[3] Goode, G. B. The Secretaries. *Smithsonian Institution, 1846–1896*, p. 135.

give Morse his support, knowing that the American inventor had a better telegraph system than either Wheatstone and Cooke in England or Steinheil in Germany. The former were seeking patent rights in this country which, if granted, might handicap Morse in proceeding with his project.

Morse and Henry continued their relations on the friendliest terms. In June of this same year (1842) the former sent his friend twelve biscuit cups with which a small battery of great power might be made. Then Henry was able to do the inventor another service which spared him from the embarrassment arising from an imperfect knowledge of the principles of electricity. Morse had received his subsidy from Congress. He was in process of laying the cable when an unexpected defect appeared.

The line was being laid in a lead pipe. Before any great progress had been made, tests showed that faulty insulation would make the whole scheme impracticable. The works superintendent was instructed to tear up the line in such a way that no one might know of the failure.

"Consultations long and painful followed. The anxiety of Professor Morse was greater than at any previous hour known in the history of the invention. Some that were around him had serious apprehensions that he would stand up under the pressure." [4]

Alfred Vail, who was Morse's associate, proposed to insulate each wire separately and to attach them in clusters on overhead poles. Neither Morse nor Vail appear to have kept abreast of their subject or they would have known that Weber, Gauss, and Steinheil had already solved this problem by attaching bare copper wires to poles by means of glass insulators. Morse's construction engineer, Ezra Cornell, worried out this idea independently but could find no member of the development staff with courage to present it in opposition to Vail's scheme. Cornell went directly to Morse, who rejected his proposal.

It should be observed that Morse was so little acquainted with electrical principles that he failed to perceive that Vail's idea was as impracticable as the original lead pipe line, and that Cornell's proposal was so sensible that it has never been superseded since it was first put to use. Morse later explained to Cornell how he came to adopt it. Cornell thus describes the incident:

[4] Prime, S. I. *Life of S. B. Morse*, 1875.

THE TELEGRAPH CONTROVERSY

Professor Morse gave preference to Mr. Vail's plan, and started for New York to get the fixtures, directing me to get the wire ready for use and to arrange for setting the poles. At the end of a week Professor Morse returned from New York and came to the shop where I was at work, and said he wanted to provide the insulators for putting the wires on the poles upon the plan I had suggested; to which I responded: "How is that Professor? I thought you had decided to use Mr. Vail's plan." Professor Morse replied: "Yes, I did so decide, and on my way to New York, where I went to order the fixtures, I stopped at Princeton and called on my old friend Professor Henry, who inquired how I was getting along with my Telegraph. I explained to him the insulation of the pipes, and stated that I had decided to place the wires on poles in the air. He then inquired how I proposed to insulate the wires when they were attached to the poles. I showed him the model I had of Mr. Vail's plan, and he said: 'It will not do; you will meet the same difficulty you had in the pipes.' I then explained to him your plan which he said would answer."[5]

With this additional proof of good will, which saved Morse no small outlay and an incalculable amount of humiliation at a critical time, relations between the two men appeared to be more closely cemented, and no minor irritation could seem to threaten the friendly feeling between them. However, minor irritants did appear, and of the sort which, when neglected, can grow into serious complaints.

The first offender was Alfred Vail, the close associate of Morse in the development of the telegraph. In 1845 Vail published a history of the electromagnetic telegraph in America,[6] a work in which no mention was made of Henry's contribution to the subject. Henry could not conceive how Morse should be ignorant of the contents of Vail's book, nor could he understand how the omission of his name could have occurred when the author was so fully informed upon Morse's indebtedness to Henry. The real discoverer of the telegraph had the right to feel slighted.

When mutual acquaintances (notably Professors Gale and Ellert) informed him that Henry felt slighted by the omission of his name,

---
[5] Morse, E. L. *Samuel Morse: his letters and journals.* Vol. II, p. 214.
[6] Vail, A. N. *The American Electro-magnetic Telegraph.* Philadelphia, 1845.

Morse wrote a note to Henry asking that he be given a hearing on the matter. He explained that he was entirely innocent, and he undertook to remove any cause of annoyance. To this note Henry attached the comment: "This is a very good letter, and I think it probable that Prof. M. intended to do me full justice, but when he came to learn the true state of the case he had not the magnanimity to do that which was right."

Whatever Morse's intentions may have been, he dictated a letter which he persuaded Vail to sign, in which the author of the offence pleaded that he had offended through ignorance, since he had been unable to ascertain the nature of Henry's discoveries. Would the professor kindly oblige him with an account of these?

This lame pretence to ignorance annoyed Henry intensely. Punctilious to the smallest degree in correspondence, he did not even acknowledge receipt of Vail's letter. Later, when he came to learn that the letter had been written at Morse's dictation, he regarded the request for information as a piece of impertinence.

However, the two men met again on amicable terms and Morse blandly admitted Vail's injustice. He gave a promise that a second edition of the book would give Henry full credit for having discovered the principle of the electric telegraph.

When the book was reissued in the following year it bore no corrections. Henry's work was still ignored, his name unmentioned.

It has been advanced in Morse's defense by his son, that he was in no sense responsible for this repeated omission. The second edition of the book is alleged to have been a reprint from the original plates with the date changed from 1845 to 1847. Henry was not aware of this, but even if he had been fully acquainted with the fact it would scarcely have diverted his displeasure from Vail, who had consented to a reprint of a text after he had apologized for its injustice. Nor could Henry entirely excuse Morse from blame for the act of his friend and associate. Some of the displeasure directed at Vail fell obliquely upon him.

Then Morse had his turn at feeling offended, although with less reason. The affront was probably intensified by the fact that it arose from the suppression of what he intended to be his overdue acknowledgment of obligation to Henry.

Professor Sears C. Walker of the Coast Survey had been engaged in writing an official report upon the value of the telegraph in geodetic survey. In order to obtain authentic material upon the development of the telegraphic system in this country he called a conference in his office which was attended by Morse, Gale, and Henry. During the course of this meeting Morse made the extraordinary statement that until he had read Henry's paper in 1847 he was unaware that the professor, while at Albany, had preceded him by two years in the discovery of the telegraph.

Leonard Gale interposed to remind him that he had introduced him to Henry's work in 1837. To this correction Morse offered no comment.

When Walker had completed his report, he submitted the manuscript to Morse, giving the inventor the opportunity to make corrections before it went to the printer. Walker had his own ideas on the value of the early work performed by Henry, and he made generous appreciation of it in his report. Perceiving that Walker could not be restrained from expressing his conviction that Henry was the real discoverer of the telegraph and had made the first crude instrument for transmitting electrical impulses to a distance, Morse wrote him a letter in which he said: "I have now the long-wished for opportunity to do justice publicly to Henry's discovery upon the telegraph. I should like to see him, however, and learn definitely what he claims to have discovered. I will then prepare a paper to be appended and published as a note, if you see fit, to your report." [7]

Whether he made any effort to consult with Henry is not known, but it seems very doubtful from the note he sent to Walker, which reads:

> The allusion you make to the helix of the soft iron magnet prepared after the manner first pointed out by Professor Henry, gives me an opportunity of which I gladly avail myself, to say that I think justice has not hitherto been done to Professor Henry, either in Europe or in this country, for the discovery of a scientific fact, which in its bearings on the telegraphs, whether of the magnetic needle or the electro-magnetic order, is of the greatest importance. While, therefore, I claim to be the first to propose the use of the electro-magnet for telegraphic purposes, and the first to construct a telegraph on the basis of the electro-magnet, yet to Professor Henry is unquestionably due the honor

[7] *Smithsonian Institution Report for 1857,* p. 91.

of the discovery of a fact in science which proves the practicability of exciting magnetism through a long coil or at a distance, either to deflect a needle or to magnetize soft iron.

It will not be argued that this tribute erred on the side of overindulgence in praise, since what was claimed for self was manifestly exaggerated and what was attributed to Henry was reduced to the minimum. No mention was made of the obstacles over which Henry had generously helped him. No hint was offered that, had Henry covered his discoveries with patents, all that would have remained of Morse's telegraph was the alphabetic code.

In making his grudging concession to Henry, Morse had left publication to Walker's discretion, it was to be appended to his report "if you see fit." As the situation had been fully exposed to him in the presence of the principals and the neutral Gale, Walker exercised his editorial prerogative and "killed" what he judged to be a deliberate misrepresentation of the facts.

This disregard for his pontifical utterance aroused the ire of the inventor, who was rapidly becoming susceptible to any suggestion that Henry was an enemy seeking to rob him of the claims to have been the first to conceive the telegraph.

Thus the two men were made resentful of each other by actions which neither controlled.

The characters of the two men were responsible for moulding the subsequent course of events. It was unfortunate that both Morse and Henry were sensitive, each possessing qualities quite different from those of the other, unlike but complementary. Had there been an equal distribution of good will, a reconciliation might have been effected. Nothing had been done that was irreparable, nor had anything been said that could not have been explained away at a chance meeting on the street.

Henry, who as yet knew nothing of Walker's editorial action which Morse regarded as an unpardonable slight, continued to pursue the course of his life as tranquilly as affairs at the Smithsonian permitted, shunning publicity, and finding contentment in his work. Morse, on the other hand, was undergoing the difficult process of adjustment in which a man's character undergoes a severe trial. It was the era of quickly and easily made fortunes. It was a generation which built up great structures of material luxuries to fill the empty places of the spirit. The nation was becoming money mad. Men went to bed to dream of arising to a dawn of riches. To Morse, the dream had come true.

On the surface he appeared to be the typical successful business man. His leap from obscurity to fame, from poverty to wealth, had been dazzling. It is small wonder he began to feel the springs of vanity stirring within his being. He revelled in the public admiration that came with recognition of his achievement; he enjoyed basking in the splendor of the great men into whose society he was thrown. But he began to create unhappiness for himself by formulating the doctrine that he, and he alone, was responsible for the wonder he had wrought in building the telegraph system.

Whatever he may have said in private conversation and in personal letters to express his gratitude for aid, he never let slip in public any expression that would convey the impression that nothing remained of his original telegraph. His first conception had been completely discarded in favor of devices suggested to him by Gale, Henry, Vail, and Cornell. Even his claim to have invented the code associated with his name has been disputed.[8]

Morse's personal contribution to the telegraph system must not be overlooked. He made it practical by his perseverance. He displayed uncommon judgement in yielding to the advice of wiser men like Gale and Vail, who filled in the gaps in his knowledge. But it cannot be denied that it was not Morse's knowledge of electromagnetism which made his success so remarkable, so much as his persistence in the face of obstacles. For that he cannot be praised too highly. Many another man would have abandoned the project, but Morse fought his way over one obstacle after another.

His beginning was modest enough. In addressing Congress for a subsidy he did not claim to have invented anything fundamental. "The chief merit . . . is that of combining together things and inventions already existing, as to produce a result never before obtained." [9]

Even this modest claim is not above question, for Morse had seen Henry's apparatus in operation at Princeton.

Had he been content to restrain his claims within this moderate limitation, no great harm would have been done, but it is character-

---

[8] Pope, F. American inventors of the telegraph. *Century Magazine*. April, 1888.
[9] Morse, S. F. B. *Memorial to Congress*. January 30, 1849.

istic of determined people that they are restricted in their vision. They look straight ahead with eyes so steady that they perceive no side issues. Trouble was bound to arise when Morse's obsession swelled so that he would admit no one to share in his honor. His biographer says: "He believed himself an instrument employed by Heaven to achieve a great result, and having accomplished it, he claimed simply to be the original and only instrument by which that result could be reached." [10] This attitude reached culmination after the courts had decided in his favor when he brought suit against those who infringed upon his claims.

A business success in those unstable days inevitably meant the uprising of a crop of imitators. Unscrupulous men were always at hand, seeking to extract a profit from another's dearly won experience. Some were satisfied with flagrant imitation, others advanced shadowy claims to have had a share in inspiring an integral portion of the enterprise. Morse encountered his full share of these competitors and impostors. When legitimate claimants to the use of the fundamental principles of electromagnetism made their appearance he treated them as though they belonged to the impostor contingent.

He was opposed by men who were sincere in their conviction that they, too, had a right to make use of the established laws of physics as applied to the telegraph. Henry was in agreement with these men and, had he wished, he could have swept aside Morse's exclusive claims. Because the courts upheld his claims, Morse was strengthened in his faith that he had discovered all that pertained to the telegraph. He hurled fierce Biblical denunciations upon the heads of all who disputed his claims, and arrived at the point where he described Henry's work as "jackdaw dreams," although he was careful to address the remark to his confidant Alfred Vail.[11]

The only person to whom he made any concession was Professor Leonard Gale, who had introduced him to Henry's work. Gale's assistance was so imperative in the early stages that he was promised one quarter share in the profits of the enterprise, but Morse later bought him out.

Had Henry been left to his own inclination he would have held aloof from all the law suits in which Morse became entangled. He

[10] Prime. *Op. cit.,* p. 494.
[11] Letter dated Dec. 7, 1853, in Vail, J. C. *Early history of the Electromagnetic Telegraph,* 1914, p. 23.

would not have volunteered to mount the witness stand to contest Morse's claims or to support those of his rivals. Samuel F. Chase, who later became Chief Justice of the Supreme Court, was counsel in some of these suits and he is our authority for saying that the only way to obtain Henry's testimony was to issue him a subpoena and to submit to him for answer a group of written questions.

Morse's son and editor of his letters and journals, while conceding that Henry's testimony was given with reluctance, asserts that it was tinged with bitterness "caused by the failure of Vail to do him justice and his apparent conviction that Morse was disingenuous." An unprejudiced observer will experience difficulty in agreeing with this verdict. Chief Justice Chase may be quoted once more for having declared of Henry "that nothing in his testimony or his manner of testifying suggested to me the idea that he was animated by any desire to arrogate undue credit to himself or to detract from the due claims of Morse." [12]

Henry was first called in the case of *Morse vs. O'Reilly,* and in testimony given at Boston, September 1847, declared that the instrument patented by Morse in his first claim had been known to Sir Charles Wheatstone and himself. He said that Wheatstone and he had earlier knowledge of the use of combined circuits; that Wheatstone, after their meeting in 1837, had employed local circuits and the principles of electromagnetism to operate his needle telegraph. He also testified that the battery used by Morse had not been invented by him; that Steinheil had made prior use of insulated wires, and as a result of this successful application Henry had recommended their adoption to Morse. He added that Steinheil and others had employed ground circuits before Morse had done so.

Henry generously praised Morse for having made a practical system. He said: "I thought his plan was better than any with which I had been made acquainted in Europe; I became interested in him and instead of interfering with his application to Congress, I subsequently gave him a certificate in the form of a letter, stating my confidence in the practicability of the electro-magnetic telegraph, and my belief that the form proposed by himself was the best that had been published."

Later Henry was called upon twice more to testify by other inventors who were successfully sued by Morse, once in favor of Bain, who

[12] *Smithsonian Institution Report for 1857,* p. 90.

had invented a "chemical" telegraph, and again for House, who had invented a printing telegraph. His testimony was much the same as that given in the O'Reilly case. He was careful to differentiate between discovery and invention, allowing to Morse full credit for the practical work he had performed.

It was not for Henry to decide whether or not the disclosure of the principles offered equal rights to all in their application; that was the prerogative of the judges. He was called upon to recite historical facts that were known to other physicists, and which could be verified by anyone who cared to consult the technical literature. As time went on there was nothing to retract nor anything to add to his evidence.

Such was not the case with Morse's expert witness. Shortly after the conclusion of the O'Reilly case, Leonard Gale complained in a letter to Morse that Henry was as frigid as a polar bear toward him. There was some reason for this change in attitude.

In his testimony in that case Gale had made a blunder which can only be attributed to the diligent coaching he had received from Morse's legal counsel. He must have been so thoroughly drilled in his answers, so indoctrinated with the impression he was to convey, that he became confused. Otherwise he would never have testified, as he did, that the first time he saw Morse's telegraph it was operating satisfactorily. This was to ignore the reason why Morse had sought his advice, and why he had suggested that Morse study Henry's paper describing the intensity battery and magnet.

Gale later corrected this error. Nevertheless, his testimony in the trial increased Henry's resentment against the Morse group, who seemed intent upon depriving him of every shred of credit for the basic work.

These law suits aroused fierce controversies in the American newspapers, several of whom depended upon O'Reilly's telegraph for gathering their news. The suits were discussed in language that is now heard only in street corner disputes, and must have been wounding to the sensitive principals. To make matters worse, a great deal of interest in the suits was aroused in England, where national sentiment awarded priority to Wheatstone's telegraph. In that country everything that served to detract from Morse's claims was welcomed as supporting those of Wheatstone.

The Morse group did not grasp the significance of this favoritism.

All they perceived were the hands stretched out to snatch the laurels from Morse's brow, and that Henry was aiding and abetting this outrage. The dispassionate judgement both in England and America was in favor of Henry as the man who had made the telegraph possible, but whose claims were being overlooked in favor of the man who had materialized his ideas.

Among the sycophants and busybodies who attached themselves to Morse after his success was assured was a "Colonel" Taliafero P. Shaffner, whose career was as histrionic as his name. He was forever inspired by dreams of great wealth which were inevitably followed by the reality of disaster in his numerous occupations. For the moment, he was publishing the short-lived *Telegraph Companion*. With the journalist's eye for a "scoop" which would assist in swelling the circulation of his journal, Shaffner sought to inflame Morse's irritation against Henry.

He began to supply Morse with misrepresentations of what the European newspapers were saying about the damaging effect of Henry's testimony in the O'Reilly case. Morse thanked Shaffner for bringing these comments to his notice, since they had escaped his attention. If the newspaper editor's intention was to goad Morse into making an attack upon Henry, he met with complete success.

But Morse had to tread warily. He had no wish to provoke a controversy in which he might incur the full extent of Henry's wrath. His patent was about to expire, and he wished to have it extended. The decisions of the courts had all been in his favor; why expose himself to the risk of losing the privileges of an extension by forcing Henry to pronounce the words which would deprive him of all protection? He chose to remain silent for the time being.

Charles Mason, the Commissioner of Patents, was by no means convinced that Morse was entitled to an extension of his patent. He was of opinion that the claims were expressed in terms covering such a broad field as to stifle all improvement in the telegraph except by Morse. He decided to consult men who were qualified to express expert opinion on the matter. Among those whom he consulted was Joseph Henry. When the time came for Henry to defend himself against attack, he asked Mason to write a letter which would reveal his attitude toward Morse in the application for an extension of the patent. Mason replied:

Some two years ago, when an application was made for an extension of Professor Morse's patent, I was for some time in doubt as to the propriety of making that extension. Under these circumstances I consulted several persons, and among others with yourself, with a view particularly to ascertain the amount of the invention fairly due to Professor Morse.

The result of my inquiries was such as to induce me to grant the extension. I will further say that this was in accordance with your express recommendation and that I was probably more influenced by this recommendation and the information I obtained from you, than by any other circumstance in coming to that conclusion.

This letter was published in the Smithsonian Report for 1857 together with other information upon the controversy, but it is not mentioned by any of Morse's biographers. It could not very well be given wide circulation by those who wished to charge Henry with being animated by bitter feelings toward Morse. The letter is conclusive evidence that Henry did not permit his grievances against Morse, Vail, and Gale to influence his judgement upon the merits of Morse's telegraph. Upon that subject Henry's opinions underwent no change.

Now that he had received the extension of his patent, Morse was free to deal with Henry.

Upon his own admission to Shaffner, Morse had been unaware that anyone in Europe had questioned his right to complete empire over the telegraph. The principal offense of the foreign journalists does not appear to have been anything more grave than their contention that Henry was the real discoverer of the principles of the telegraph, but that he had not received recognition for his accomplishment. This was inadmissible to Morse. He began the preparation of his claim to be the sole discoverer and inventor of the electric telegraph.

This appeared in 1855 in Shaffner's *Telegraph Companion* under the title of "Defense against the injurious deductions drawn from the deposition of Professor Henry." The "Defence" filled the entire ninety-six pages of the issue. Shaffner had secured his "scoop."

Had it been what it purported to be, Morse's defence could have aroused no dispute. But instead of being a correction of erroneous inferences, this literary effusion of Morse was an unveiled attack upon Henry's ability as a scientist and an effort to destroy his credibility as an expert witness. Near the beginning of the "Defence" he said that he proposed:

> First, certainly [to] show that I have not only manifested every disposition to give due credit to Professor Henry, but under the hasty impression he deserved credit for discoveries in science bearing upon the telegraph, I did actually give him a degree of credit to which subsequent research has proven him to be not entitled.
>
> Second, I shall show that I am not indebted to him for any discovery in science bearing upon the telegraph; and that all the discoveries of principles having this bearing, were made, not by Professor Henry, but by others, and prior to any experiments of Professor Henry in the science of electro-magnetism.

Common sense should have warned Morse that his first proposal was an ungracious attitude to adopt toward the man who had freely given him counsel, support, and encouragement. It was bound to reflect upon him as base ingratitude. His second proposal was beyond human power. The published evidence flatly contradicted all he had to say.

Morse had many fine qualities, but modesty was not among them. His display of ingratitude was not restricted to relations with Henry. Nevertheless, it would be futile to deny that his sublime faith in himself and his invention, and his dogged perseverance at considerable self-sacrifice deserved appropriate reward. Yet it is equally clear that because of technical ignorance he was heavily indebted to Gale, Vail, and Henry. Candor should have caused him to admit this, but to none would he extend credit.

He obstinately persisted in the foolish contention that the idea of the electric telegraph came in perfect form when he conceived it in 1832 on the *Sully,* while returning from England. In his own deposition in the Bain case, he was compelled to admit that this scheme was impracticable. He was merely following in the footsteps of the luckless Barlow. He would not have advanced beyond that stage had not Chilton and Gale advised him to adopt the principle of Henry's intensity battery. In his "Defence" he repudiated this advice. In fact, there is not in the whole of Morse's writings and speeches a single remark that would indicate he ever learned the principle of intensity and quantity circuits, so vital to the telegraph.

From a man who had twice endeavored to obtain from Henry an account of what he had discovered, Morse displayed in his "Defence" an uncanny intimacy with Henry's papers. In his first contribution to Silliman's *Journal* in 1831, describing his first magnets, Henry had stated "the principal object in these experiments was to produce the greatest magnetic force with the smallest quantity of magnetism." Morse eagerly seized upon this simple and early affirmation as proof that Henry was not engaged in any experiments relating to the transmission of electrical impulses to a distance before Morse conceived his own, unworkable, scheme in 1832.

In this same paper Henry carelessly alluded to "Mr. Barlow's project of forming an electro-magnetic telegraph," when he actually was proving that Barlow's announcement *against* such a project was erroneous. The context is sufficient to prove that this was a textual slip committed in the haste of writing the paper. Morse seized upon it as conclusive proof that while he was developing the telegraph (which he judiciously refrains from mentioning was impracticable) Henry was very ill-informed upon that subject.

Morse's vitriolic "Defence" has never been satisfactorily explained away by his biographers. It stamps him, as nothing else could have done, as consumed by self-sufficiency. He was so deeply concerned with his own approbation that he was blinded to all sense of justice to others. His main object of attack was Henry, and truth was distorted to castigate him for presuming to assert that other men had had a hand in performing "what God hath wrought."

When Morse's charges are balanced against incontrovertible facts, how is one to account for his action in making them public? It is always a dangerous course to impugn a man's candor. Perhaps a later writer who investigated the charges comes nearer the truth when he supposes Morse to have been temporarily unbalanced.[13]

Although not one who spread the peacock feathers, as Morse delighted in doing, Henry was proud and sensitive; he insisted upon receiving his due. He was deeply wounded by Morse's attack. Had he been a private individual it is unlikely he would have taken any steps to controvert the misrepresentations. He would, most probably,

[13] Harlow, A. F. *Old Wires and New Waves*, 1936.

THE TELEGRAPH CONTROVERSY

have allowed the evidence to speak for itself, for he would assume that those who knew the circumstances would rectify the errors before history.

But Henry was no longer a private individual with nothing but his personal feelings to consider; he was the chief executive officer of a public institution, whose honor and professional reputation had been challenged. Coming, as it did, immediately after the struggle over his staff relations and the dispute about his library policy, the attack made him acutely sensitive. Morse had presented the more unscrupulous opponents of that policy with ammunition to fire at the Smithsonian's Secretary.

However, Henry acted with the customary caution. Instead of hurrying into print with a refutation of the charges against him, he thought first of his responsibility to the institution under his charge. He asked Chief Justice Taney, who was then Chancellor of the Smithsonian Institution, for advice on the wisest course to follow. Taney said that he had read Morse's "Defence" but had given it very little attention, as he supposed it to have been inspired by one of the opponents to Henry's Smithsonian policy. He recommended that the excitement and prejudices which had flamed during the past year be allowed to subside before any action was taken. Henry was advised to assemble whatever material he considered necessary to refute the charges, and to publish it at his own discretion.

The year 1856 was allowed to pass without any public answer to Morse's attack. Then, Henry displayed his wisdom by choosing a method of confutation which abundantly cleared his honor, and established beyond doubt his reputation as a scientist, without having to enter into any verbal conflict with his accuser.

Disregarding Morse's method of appealing to an uninstructed public, Henry addressed himself to the Board of Regents and requested the appointment of a committee to investigate his claims to the original scientific work in relation to the electric telegraph, and to exonerate him from the charges and implications contained in Morse's "Defence."

The Regents acceded to his request, and appointed a committee consisting of Senators James M. Mason, James A. Pierce, President Cornelius C. Felton of Harvard College, and Stephen A. Douglas. To them, Henry submitted a bald statement in which he recited the recorded facts of his work. In order to remove any impression that he had been animated by rancor toward Morse, or advanced any claims to achievements to which he was not entitled, he supplemented his statement with testimoney from competent witnesses.

He requested the declaration from Justice Chase, from which we have quoted, to prove that he had volunteered no evidence likely to hamper Morse in his suits, nor had he been moved by jealousy or animosity toward the inventor. Chase freed him from any such imputations.

Mason, the Commissioner of Patents, was invited to make a statement on Henry's attitude toward Morse at the time the inventor was applying for the extension of his patent, and he responded with the letter we have quoted.

The most injurious of these supplementary documents came from Professor Gale. Henry did not doubt Gale's veracity. He was convinced that the erroneous statement on the witness-stand had been made in confusion created by the suggestions and ambiguities offered by the lawyers in their questions to witnesses. Accordingly, Henry had written to Gale begging him "to state definitely the condition of the invention when he first saw the apparatus in the winter of 1836."

Gale's reply left nothing to be desired. He wrote:

> ... The sparceness of the wires in the magnet coils and the use of the single cup battery were to me, on the first look at the instrument, obvious marks of defects, and I accordingly suggested to the Professor, without giving my reasons for doing so, that a battery of many pairs should be substituted for that of a single pair, and that the coil on each arm of the magnet should be increased to many hundred turns each; which experiment, if I remember aright, was made on the same day with a battery and wire on hand, furnished I believe by myself, and it was found that while the original arrangement would only send the electric current through a few feet of wire, say 15 to 40, the modified arrangement would send it through as many hundred. Although I gave no reasons at the time to Professor Morse for the suggestions I had proposed in modifying the arrangements of the machine, I did so afterwards, and referred in my explanations to the paper of Professor Henry, in the 19th volume of the *American Journal of Science,* page 400 and onward.
>
> At the time I gave the suggestions above named, Professor

Morse was not familiar with the then existing state of the science of electro-magnetism. Had he been so, or had he read and appreciated the paper by Henry the suggestions made by me would naturally have occurred to his mind as they did to my own.... Professor Morse expressed great surprise at the contents of the paper when I showed it to him, but especially at the remarks on Dr. Barlow's results respecting telegraphing, which were new to him, and he stated at the time that he was not aware that anyone had even conceived the idea of using the magnet for such purposes.

One wonders what Morse thought when he read this letter from his friend, which irrefutably destroyed his claims to have been the original pioneer of the telegraph, or to have developed it without the aid of any other person.

Finally, to prove that he had experimented with a crude but workable telegraph instrument, embodying the fundamentals which Morse adopted, Henry appealed for evidence from Professor James Hall, a former president of the American Association for the Advancement of Science. In reply, Hall declared that he had called, with a letter of introduction from Amos Eaton, upon Henry at the Albany Academy in August 1832, and had been shown the little bell with its mile long wire connection to a galvanic apparatus. He recalled Henry's remark that bells could be made to ring at a distance of several miles by such an arrangement. Since the development of the telegraph by Morse, this visit to Albany had frequently recurred to his mind, and he had recited the facts of the visit to many who had spoken in his hearing of the marvel Morse had accomplished.

In the presentation of his case, Henry made no recriminations, offered nothing to belittle or to disparage what Morse had done. The facts were arrayed and permitted to speak for themselves.

The special committee appointed by the Regents completely exonerated Henry from the charges levelled against him. The members forthrightly condemned Morse's "Defence" in these words:

> The first thing which strikes the reader of the article is that its title is a misnomer. It is simply an assault upon Professor Henry; an attempt to disparage his character; to deprive him of his honors as a scientific discoverer; to impeach his credibility

as a witness and his integrity as a man. It is a disingenuous piece of sophisticated argument, such as an unscrupulous advocate might employ to pervert the truth, misrepresent the facts, and misinterpret the language in which the facts belonging to the other side of the case are stated.[14]

The Regents declared further that "Mr. Morse's charges not only remain unproved but they are positively disproved." This cleared Henry of all the charges against his honor, his integrity, and his scientific reputation much more effectively than Henry could have done by an appeal to any other tribunal. Certainly no public statement above his own signature would have carried the weight of the opinion signed by these Regents. In the outcome he suffered much less from the attack than did the man who made it. It lost Henry no friends.

Having had his character vindicated, Henry permitted the unpleasant episode to slip from his mind. He never again alluded to it.

Morse was not so ready to accept a verdict which questioned his sole right to the telegraph and everything concerned with it. He was obdurate in refusing to concede that Henry had contributed to the accomplishment. Ten years later, when writing his report on the electrical section of the Paris Exposition, in the capacity of Commissioner of the United States, he contrived to repeat Vail's blunder of omitting Henry's name when writing the history of the electric telegraph.

The general editor of the reports was Professor W. P. Blake, who instantly perceived the omission. In drawing Morse's attention to it he said: "Frankly I am pained not to see the name of Henry there associated with those of Arago and Sturgeon, for it is known and generally conceded by men of science that his researches and experiments and the results which he reached were of radical importance and value, and that they deservedly ranked with those of Ampère, Arago, and Sturgeon."

Blake was well aware of the estrangement between Morse and Henry. Like everyone else acquainted with its cause, he deplored it, and proposed to act as mediator in bringing the two men to an understanding. In making the proposal to Morse he was careful to announce that he had not consulted Henry upon his readiness to receive any

[14] Special Committee's Report. *Smithsonian Institution Report for 1857*, p. 98.

approach. With the magnanimity of one who forgives an enemy for the injury about to be inflicted upon him, Morse gave his consent to a meeting. But Blake would not approach Henry until Morse had made fitting retribution in his report. He was not going to attempt a reconciliation knowing that Morse was about to display further ill-will.

So, having failed to obtain from Morse in writing any tribute to Henry's contribution to the science of electromagnetism, Blake paid Morse a personal visit and urged his host to make the necessary concession. Finally, Morse consented to include in his statement the grudging paragraph: "In more recent papers, first published in 1857 it appears that Professor Henry demonstrated before his pupils the practicability of ringing a bell by means of electro-magnetism at a distance."

This erroneous and reluctant concession was as far as Morse would go, but it was far from satisfying Blake. He did not accept the ungracious admission as a satisfactory basis for a reconciliation between Morse and Henry. Despairing of making any further progress, he abandoned his efforts.

E. L. Morse, in the publication of his father's letters and journals makes a curious comment on this incident when he writes: [15]

> Whether Professor Blake was satisfied with this change in the original manuscript is not recorded, Morse evidently thought he had made the *amende honorable,* but Henry, coldly proud man that he was, still held aloof from reconciliation, for I have been informed that he refused to be present at the memorial services held in Washington after the death of Morse.

The words "honor" and "pride" assume a strangely unfamiliar connotation when applied to Morse's attitude to his victim.

In arriving at a final judgement in the dispute between Morse and Henry, let us quote the words of an unprejudiced observer, written several years after the controversy had ended and both principals were in their graves.

It was only by Henry's discoveries that the electro-magnetic telegraph of Morse became possible, and Morse himself, before

[15] Vol. II, p. 479.

he became involved in patent litigation, freely acknowledged his indebtedness to Henry. But Professor Henry, long before Morse's telegraph came before the world, had suggested the application of his electro-magnets to telegraphy, and even had constructed a form of bell telegraph for experimental purposes which answered remarkably well. Henry, however, had for his object "the advancement of science, without any special or immediate reference to its application to the wants of life or useful purposes in the arts. . . ." He gave freely to the world the results of his researches, and others devoted themselves to the practical application of the principles which he discovered. Of these were not only Morse in America, but Wheatstone and Cooke in this country. It has been amply demonstrated that these inventions were at a standstill in the early part of 1837 for want of the means of producing a strong current at the receiving station. Although Henry had clearly shown the advantages of employing closely wound coils of fine wire in 1831, Wheatstone knew nothing apparently of this, and remained in ignorance until April, 1837, when he was enlightened by Professor Henry himself. We are firmly convinced Henry did more for the advancement of the telegraph than has yet been adequately acknowledged.[16]

[16] *Electrical Review*. London. August 12, 1887. Vol. 21, p. 162.

THE RIPENING YEARS

ALTHOUGH the future was not to be all sunshine without shadows, the stormy period of Henry's life was over. There was to be another struggle, but this was to be a private affair. Henry had to settle for himself the problem of what his attitude should be toward the intransigeant Southern states.

He did not keep his nose so close to the scientific grindstone that he was oblivious to what was happening to the nation. If nothing but vague whispers of rebellion darkened the bleak austerity of President Buchanan's parties to which Henry was invited, more substantial evidences of the forthcoming calamity were walking abroad. The premonitions of approaching battle rumbled through the quiet corridors of the Smithsonian, for many of the Regents were prominent political figures, and not all were of the same opinion or party.

The one consuming passion of Henry's life now was the welfare of the Institution, and he prepared to work for this in any political circle. His friendships were not influenced by party views. One of his earliest acquaintances on coming to Washington was with Andrew Johnson, then holding a seat in the House of Representatives. Johnson was frank enough to admit that he did not favor the Smithsonian policies, whereon Henry undertook his conversion, and when Johnson succeeded to the presidency, Henry was able to write that he had successfully cultivated the acquaintance until the new President's prejudices had been overcome, and he could be counted upon as one of the ardent supporters of the Institution.

Henry had also formed a friendship with Jefferson Davis while the latter was Secretary of War during 1853–1857. The two men were probably brought together by Bache, who had known Davis during their West Point days. Davis had a keen appreciation of Henry's intellectual gifts and was among the first of the government department heads to invite his cooperation in the solution of a technical problem relating to the properties of a new explosive. Intercourse was not restricted to their official association. Their relations became cordial, and continued on that plane after Davis vacated his cabinet seat and entered the Senate. Professor and Mrs. Henry were frequent guests at the Davis home at F and Fourteenth Streets.

Intimacy with the southern leader was not necessary to acquaint any resident of Washington with the extent of the rift between the North and the South. Washington society had developed a taste for lavish entertainment, and the tall figure of the Professor was to be seen in attendance upon the belles who shone beneath the chandeliers. He had that native courtliness which was designed to secure for him a place in society, and he enjoyed meeting people, but he never permitted himself to become the slave to society that his fellow-scientist Humphrey Davy became.

Washington society in those days was controlled by women. In the atmosphere they created, where poetry and tragedy, but not humor, were encouraged, Henry could make himself at home with a graciousness that assured a welcome.

Whoever listened to the conversations in a Washington drawing-room during the year 1859, after John Brown's raid on Harper's Ferry, could not remain unconscious of the coming conflict. It was a momentous year, and the talk was not confined to low mutterings of armed rebellion. The undertones of war must have been subdued when Henry went to the British embassy to shake the hand of Baron Renfrew, as the heir to the British throne preferred to be called during an unofficial visit to this country.

An atmosphere far removed from grim visaged war prevailed at the reception given to the Japanese ambassadors in their gay costumes. Henry was deeply engaged with the problems of exploration at the time, and there was much to interest an inquiring mind in these people from an unknown country. Henry had talked over the projected visit of this mission with President Buchanan, and had been consulted upon the choice of a suitable person to attend to the care of its members. He must have made a favorable impression upon the visitors, for he was presented with a handsome dinner service by them, which was bequeathed to the Smithsonian by his last surviving daughter.

But these were mere episodes temporarily distracting attention from the absorbing topic of secession and a possible war. Everyone in a frontier town like Washington had to decide upon which side his sympathies lay. Henry was first and foremost a scientist. He had never mingled in politics, so that when he was faced with the necessity of choosing a policy, it seemed that he would have chosen, like the majority of citizens, to follow the custom in his own state and to console himself with the reflection that, by so doing, he was in good company. This he did not do. He seems to have been guided more by strong personalities.

His nearest approach to a statement of political principles at the time is given in a letter to Asa Gray.

> Smithsonian Institution,
> August 14th, 1860
>
> The good people of the North do not appear to believe in the fact of the danger in which we are now placed. I think it more than probable that our union is doomed to suffer the fate of all governments. The struggle of life must produce its effects with us as it has with all the world of past history. Every year the number of persons who adopt politics as a profession is increasing, the class must therefore deteriorate in talents, requirements, and morality. The struggle for office must constantly increase in intensity, and as under one organization the number of offices cannot be changed the tendency will be to separate us into several governments each with a president and a corps of officials.
>
> . . . If the North will suffer the negro question to remain undisturbed the whole matter will in due time be settled by the law of population and the conflict of races. Labor from the North, as it is hampered by the increase of laborers must be gradually extended into the South until it is stopped by the heat of the sun.
>
> There are parts of our country which cannot be worked by the white man, and this must be cultivated by the negro or not at all. I have little hope that the black man can ever be civilized unless by selection in the course of geological periods, but I would make the experiment on a grand scale and expend millions under the direction of the colonization society in establishing an empire

in Africa. The rulers might be the half-breeds which we could consistently furnish for many years to come.

I do not think the negro can ever exist in close approximation with the white man except in a state of slavery. The struggle of life must be most severe at the lowest point of the scale and the negro has neither the mental nor physical power in our climate, to stay long in the contest.

He was now sixty, beyond the age which experiences the uplift of the spirit traditionally associated with battle. He was of the age which sees in war the shattering of the house of life which one has built for oneself, and the despair of building another. It is a dangerous age for war, an age in which one too readily sinks into acquiescence of the tragedy, without hope or philosophy. Yet he was at first inclined to sympathize with the South in aspirations which could only lead to warlike action.

Much as one might like to think he had the discernment to perceive the elements of greatness and righteousness in President Lincoln, it must be confessed that it was not until he came under the influence of that great leader's personality that he began to see the truth.

Henry invited his old friend Torrey to visit Washington in order to hear Lincoln's inaugural address. On his arrival, Torrey found that a Mr. and Mrs. Bell, ardent Southern sympathizers, were also guests in the Henry apartment, and it was apparent that their strong feelings were reflected by their hosts. In a letter to Asa Gray, Torrey described Professor and Mrs. Henry as being bitterly opposed to the new President.[1]

However, after meeting Lincoln, Henry's views underwent a change.

In 1862 a visit to L. E. Chittenden, Register of the Treasury, on official business led to a meeting which enabled Lincoln and Henry to express their views on each other's character. When their business, which had to do with the work of the Light-house Board, was completed, Chittenden and Henry fell to talking about the President. Henry was asked for his opinion on Lincoln. He confessed that his admiration increased as their acquaintance developed. "I have lately met him five or six times. He is producing a powerful impression upon me. It increases with every interview. I think it my duty to take

[1] Rodgers. *Torrey*, p. 272.

philosophic views of men and things, but the President upsets me. If I did not resist the inclination, I might even fall in love with him."

When Chittenden expressed agreement with this sentiment, Henry was encouraged to enlarge upon his view.

> President Lincoln impresses me as a man whose honesty of purpose is transparent, who has no mental reservations, who may be said to wear his heart on his sleeve. He has been called coarse. In my interviews with him he conversed with apparent freedom, and without a trace of coarseness. He has been called ignorant. He has shown a comprehensive grasp of every subject on which he has conversed with me. His views of the present situation are somewhat novel, but seem to me unanswerable. He has read many books and remembers their contents better than I do. He is associated with men who I know are great. He impresses me as their equal, if not their superior. I desired to induce him to understand and look favorably upon a change which I wished to make in the policy of the Light House Board in a matter requiring some scientific knowledge. He professed his ignorance, or rather he ridiculed his knowledge of it, and yet he discussed it intelligently.

Henry had not finished expressing his views but, when he had arrived at this point, the door was thrown open and the President was announced.

"You have just interrupted an interesting commentary," Chittenden laughingly observed as he rose to meet his visitor.

"Do not! You will not say another word," Henry blushed in confusion. "You will mortify me excessively if you do."

Lincoln and Henry then engaged in a long discussion upon the difficulties encountered by Union vessels navigating in waters from which the Confederates had removed all lights and buoys. The latest stage of the scientist's experiments on behalf of the Light House Board had to be described. Then, when Henry had apologized for consuming so much of the President's time and had left the room, Chittenden seized the opportunity to inquire what Lincoln thought of the other man.

"I had the impression the Smithsonian was printing a great amount of useless information," Lincoln answered. "Professor Henry has convinced me of my error. It must be a grand school if it produces such thinkers as he is. He is one of the pleasantest men I have ever met; so unassuming, simple, and sincere. I wish we had a few thousand more such men."

Evidently the two men had begun with wrong impressions of each other's characters and motives, but they were both big enough to admit that they had erred. Once the battle was joined between the North and South, Henry shed all his doubts and worked with a wholehearted enthusiasm for the northern cause. Having chosen his ground, he had to decide what was to become of the Smithsonian if war should break out.

Although conscious of his responsibility for guiding the course of the Institution, he was not afraid of the future. With the light rain of April, the chestnuts on the Mall broke into flower, intimating that war or no war, nature would carry on her work. Henry must have taken this as his cue. He resolved that the Smithsonian would not curtail its activities but would continue its mission as far as was possible. Neither the flight of the inhabitants who feared a siege nor the military frenzy which seized the remaining citizens when political buncombe yielded to bayonets was to be allowed to interfere with the established routine.

In making this decision, the Secretary had overlooked some of the contingencies of war.

After Virginia seceded and Maryland temporarily fell into the hands of the rowdies, Washington was an unhappy city. Adding to the discord among friends was the confusion created when troops poured into the capital where no preparations had been made for their reception. The presence of the troops served to dispel the gloomy apprehensions of a siege. They did more; they relieved the occupants of the Smithsonian from a real siege.

From 1858, the Henry family had been occupying an apartment in the East wing of the building, which had been designed as an official residence for the Secretary. This would have been a desirable place to live if one could disregard its proximity to the open sewer, called the Canal. This offensive strip of water had to be crossed every time one entered or left the building. In one of his letters, Henry complained that "the air was redolent with the odors of stables, hospitals, and the canal."

The tumult in the capital could not flow around the building without some effect upon its tranquility. The city swarmed with underground characters drawn from most of the states in the Union. One particularly active band of robbers established headquarters in the Smithsonian grounds, from which they sallied forth at night in search of their prey. Although the residents in the Smithsonian building ventured out at night at their own peril, Henry declared that he had not been challenged by this band.

Then came the proposal that the building should be handed over to the military authorities to accommodate some of the troops who were being hurried into the capital. A meeting of Regents was hastily summoned to decide how this emergency should be met. They reached the conclusion they had no authority to *offer* the building for a use other than that intended by the founder. If the Secretary of War was of the opinion that possession of the building was a military necessity, he must assume responsibility for requisitioning it. The Regents also pointed out that, if the building was to be put to a military use, it would best serve the purpose of a hospital. The building was never requisitioned.

The nearby grounds were occupied by troops for target practice. As the garrison increased in numbers their presence was inescapable. The brazen blast of bugles, the thumping of drums, and the rattle of musketry began at dawn, continued through the day, and sometimes endured far into the night.

The great parallel sheds of the Armory Square Hospital were erected close to the Smithsonian. The problem of furnishing this hospital's primary needs gave the Smithsonian staff its first military task.

Pitiful conditions prevailed in the army hospitals during the war between the North and the South, for neither side had made preparation. After the first rush of patients, the Armory Square Hospital exhausted its supply of disinfectants, and the senior medical officer inquired of Professor Henry whether the resources of the Smithsonian could provide for his immediate needs.

Unable to furnish the disinfectants demanded, Henry set his staff to work in the chemical laboratory to produce them for future needs, and they were soon in a position to supply other hospitals, in addition to their neighbor.

A less happily inspired incident brought a certain amount of ridicule upon Henry's head.

He was intensely jealous of the Institution's good name, and this made him acutely sensitive about its political associations. He was suspicious of some individuals and organizations applying for use of the lecture hall. He was fully aware that some requests were made for the use of this hall with no other purpose in view but to create the false impression that the Smithsonian was giving its benediction to the purpose of the meeting.

With reluctance, he had given permission for the use of the hall for a series of lectures upon the anti-slavery movement. The speakers were men of the highest reputation, like Henry Ward Beecher, Horace Greeley, and Wendell Phillips. But after breathing the Washington atmosphere for fifteen years, Henry was only too well aware that the anti-slavery movement was far from popular with a strong faction. When it was apparent to him that the Smithsonian might incur some of the animosity of the pro-slavers by permitting its platform to be used by their opponents, his native caution induced him to attempt to divert ill-will from the neutral institution. He insisted, therefore, that each lecture be prefaced by an announcement read by the chairman excluding the Smithsonian from association with the ideas advanced by the speakers.

It was an unwise exercise of caution.

When the chairman read the disclaimer: "Ladies and gentlemen, I am requested by Professor Henry to announce that the Smithsonian Institution is not in any way responsible for this course of lectures," he usually followed with the further comment, "I do so with pleasure, and desire to add that the Washington Lecture Association is in no way responsible for the Smithsonian Institution."

The supplementary announcement never failed to provoke an outburst of laughter among the audience, some of whom must have been aware of Henry's intimacy with Jefferson Davis before the war. Most of the audience probably failed to understand that the unhumorous Henry stood for something above party, or even above nationality.

Accompanied by Welles, Chase, and Bates of his cabinet, Lincoln attended one of these lectures to hear Horace Greeley. The customary gale of laughter greeted the reading of the disclaimer of responsibility. Lincoln chuckled with delight. Later, on visiting Henry in his apartment, the President said, "The laugh was rather on you, Henry." [2]

Henry must have admitted ruefully that the President was right. Laughter can be a dangerous weapon. He was not made the butt of

---

[2] Quoted in Sandburg's *Abraham Lincoln: the war years.* Vol. I, p. 401.

another joke of this kind, for he never again permitted the use of the lecture hall for anything but a scientific purpose.

The towers of the Smithsonian Institution were occasionally used for testing visual signalling for the army, and when these tests were made at night, the President observed the results. The flashing of lights on the towers caused a certain amount of uneasiness among citizens who were not in the secret. In the fall of 1861, Lincoln was visited by a caller who declined to be rebuffed when told that the President was engaged with another visitor. Urging that he came upon an errand of national importance, the caller was admitted.

On entering the room where the President awaited his coming, the visitor was embarrassed to find that the chief of state was not alone. Observing the visitor's hesitation to speak, Lincoln encouraged him to proceed with the utmost freedom, as the other gentleman enjoyed his confidence.

The stranger then unfolded his story. On several occasions he had observed a light being displayed on one of the towers of the Smithsonian building for a few minutes immediately after nine o'clock. The light moved in such mysterious fashion that the observer concluded that someone within the building was flashing messages to the rebels occupying Munson's Hill.

Lincoln listened to the recital with a grave face. When it was ended, he turned to his companion with the question:

"What do you have to say to that, Professor Henry?"

Henry explained to the visitor that at nine o'clock each night it was the duty of one of his assistants to read the meteorological instruments set out on the roof of that particular tower and, as no other form of illumination had been provided, the official was compelled to carry a lighted lantern.

The abashed visitor would have fled in confusion had he not been restrained by both Lincoln's and Henry's praise for the conscientious manner in which he had discharged his duty in protecting the security of the capital. He departed with a feeling of satisfaction rather than humiliation.

After he had gone, Lincoln made no effort to maintain his gravity. It is not recorded whether Henry saw the humor in the situation.

In 1865 a defective flue caused a fire to break out which destroyed the roof and all the interior of the upper story of the main portion of the Smithsonian building, the interior of the two north towers, and also the larger south tower. The personal effects of James Smithson were almost totally lost, a valuable collection of paintings of Indians by J. M. Stanley was destroyed, but worst of all, Henry's correspondence, notes, and manuscripts were burned.

This was a cruel blow. Precious links with the past, mementoes of friendship with men who were dead and gone, all the memoranda of work done and that awaiting accomplishment, had vanished in a swirl of flame and smoke. The loss is the despair of Henry's biographers, since it draws a barrier across many passages of his life. However, Henry accepted it with calm submissiveness. "A few years ago, such a calamity would have paralyzed me for future effort," he confided to Torrey, "but in my present view of life I take it as the dispensation of a kind and wise providence, and trust that it will work to my spiritual advantage." The tone of this comment on the loss of his manuscripts, letters, and records would indicate that Henry had reached the mellow age when the mind is less sensitive to material gain and loss.

Sad though the loss of his possessions might be, it shrank into insignificance alongside the loss he suffered in the death of his only son, William Alexander. This young life had been charged with high promise during the young man's residence at Princeton. Other bereavements were to follow which must have left Henry a lonely man, in spite of his great circle of friends and acquaintances. The swift blow which struck down President Lincoln deprived him of a true friend. Close relations between the two had grown rapidly for, while they had different origins and backgrounds of education and training, both were self-made men, and neither lacked ardor in friendship.

The inventor streak in Lincoln caused him to take an interest in many of the scientific ideas advanced by practical men. He had a natural curiosity which made him desire to know the physical law or mechanical principle underlying a phenomenon that came under his notice. When he could not find the answer without assistance he would submit his problem to Henry. Little by little the acquaintance grew with the increasing frequency of Lincoln's visits or with Henry's calls at the White House as he became more deeply enlisted in the government's service.

Shortly after the close of the war, Henry lost his dearest and closest friend outside his family, Alexander Dallas Bache. No man could have asked for a worthier friend than Henry found in Bache. Their friendship had grown into close intimacy while Henry was at Princeton and Bache, great-grandson of Benjamin Franklin, was Professor of Natural Philosophy at the University of Pennsylvania. It had been joyfully renewed when Henry moved to Washington, where Bache was already established as head of the Coast Survey. It was Bache who persuaded his friend to forsake Princeton and the work of an experimental physicist in order to accept the position at the Smithsonian. As an original member of the Board of Regents, he had been a constant support and an unfailing friend during the time of Henry's most arduous labors and most acute anxieties.

Bache was not only a man of great personal charm and moral integrity; his scholarship was almost flawless, his industry unflagging. Added to these qualities was an astonishing success as a man of affairs. One is entitled to doubt whether Washington had his equal in sagacity in the conduct of affairs during the years of his residence, and we may assume that Henry had the full benefit of his knowledge of men.

Although the two men were within easy reach of one another while in the capital, they kept up a steady correspondence to supplement their conversations. After Bache's death, Henry and his wife remained the principal confidants and closest friends of his widow.

Other old friends began to pass from his company during these years between the opening of the Civil War and the settlement of its rivalries. Benjamin Silliman had died, full of honor. One who did not reach any high pinnacle of fame, but who enjoyed a close friendship with Henry, was Joseph Saxton.[3] During the period of Saxton's activity they had much in common, and when he was stricken with paralysis, Henry was a comfort to the invalid.

The war years were a turmoil which cast up many men and tore away others from Henry's circle. One of those from whom he expected great things but who failed him sorely was George Gordon Meade. For more than a decade Henry had received proofs that Captain Meade possessed unusual scientific qualifications, for the young engineer officer had been attached to the Light House Board. Later he was placed in charge of the Great Lakes Survey, where his talents had a wider field for exercise.

When Meade agitated to be placed on active duty immediately on

[3] Saxton, whom Henry met at The Franklin Institute, was the friend of many scientists in England and America. An instrument maker of exceptional talent, he is best-known for his invention of the reflecting pyrometer and his construction of standard balances used in the annual assays at the Philadelphia Mint.

THE RIPENING YEARS

the outbreak of hostilities, Henry sought to discourage him from pressing his request. Arguments pointed to prove that science could not afford to lose such a promising recruit failed to shake the soldier's resolve, whereupon Henry went to Philadelphia where he beseeched Mrs. Meade to lend her aid in preventing a step which would lead to a great loss for science. He expressed the opinion that it was sheer waste for one possessed of Meade's abilities to relinquish a brilliant future, as he expressed it, "to become food for powder." [4]

It would have been interesting to have had Henry's views upon this judgement after Meade's brilliant victory at Gettysburg, which helped materially to turn the tide of war in favor of the northern armies.

This period of Henry's ripening years has been frequently recorded as a time when he was torn from scientific pursuits to lose himself in the banalities of administration. Nothing could be further from the truth. Scientifically it was far from being a barren period, although it produced no great discovery. He was actively employed, and his activities displayed more markedly the thread of pragmatism which had run through the fabric of his work in pure science. There is a shade of atavism in Henry. He resembles somewhat the primitive "scientist" who was a composite of magician and craftsman.

His versatility during the war years was phenomenal, although the professional scientist will agree that one who has closely investigated one field of science brings into a new field a point of view which gives him special qualifications for understanding not shared by the novice.

Henry was ever ready to kindle to a fresh vision. New facts made an instant impression upon his capacious and sensitive mind. The flexibility of his mind and the breadth of his knowledge are revealed in the execution of tasks which were brought to him. To each sphere of work he seems to have carried the right complement of gifts and he finished each stage of the work so that it remained behind him like a satisfying work of art. He took up difficult tasks but he recorded their accomplishment in such modest terms that he made them look very simple. Thus both the significance and the magnitude of his work has been overlooked.

It has been too readily assumed that the bent of his mind was toward basic rather than toward utilitarian research. This misapprehension may be attributed to the nature of his discoveries in elec-

[4] Meade, G. *Life and Letters of George Gordon Meade*, 1913. Vol. I, p. 217.

tricity and magnetism, which were of the first magnitude. Nothing he produced in later life can compare with them, but some of these later accomplishments have a merit which should have rescued them from the undeserved oblivion into which they have fallen.

Henry drew a distinction between the advance of science and the progress of the practical arts, but there was no snobbery in his attitude. He added his plea to the cause most in need of support, but he never shrank from engaging in a study which had a practical purpose, and his contributions to applied science deserve recognition.

An excellent example of applied science blending with basic experimental investigation is the work he performed as a pioneer in testing materials of building construction.[5] In 1851, President Fillmore appointed a committee to examine the various marbles offered for the construction of an extension to the Capitol buildings. Henry was a member of this committee and of another which continued the work in the later stages of construction. In describing the work of the committee, he begins with an account of the problem confronting the members. "While the exterior materials of a building are to be exposed for centuries, the conclusions to be desired are to be drawn from results produced in the course of a few weeks. Besides this, in the present state of science, we do not know all the actions to which all the materials are to be subjected by nature, nor can we fully estimate the amount of those which are known."

This was the kind of problem which appealed to him. It implied a combination of talents, of theory and experiment. The existing state of knowledge offering no satisfactory solution to the problem, the members of the committee decided to institute their own standards and tests.

Small blocks of marble were put under a press to test their resistance to crushing. It was found that the bare blocks withstood twice as much crushing pressure as they did when tested by the current practice, with sheets of lead inserted between the press-plate and the marble. Naturally, Henry was not satisfied with a purely empirical result which contradicted long practice. He had to have a reason. He pointed out "in the case where rigid equable pressure is employed, as in that of thick steel plates, all parts must give way together. But in that of a *yielding* equable pressure, as in the case of the interposed lead, the

[5] Mode of Testing Building Materials. A.A.A.S. *Proceedings*. 1855. Vol. IX, p. 102; S.W. Vol. I, p. 344.

stone first gives way along the outer lines of least resistance, and the remaining pressure must be sustained by the central portion around the vertical axis of the cube."

In this same address, he also pointed out the way to progress in constructional engineering. A study of the conventional materials would leave knowledge static. He suggested that "the greater tenacity of iron and its power of resistance to crushing" made it a desirable material for the construction of buildings of more slender design with a lighter arrangement of parts.

But, while he had to share the credit for these results with his fellow members of the committee, he proclaimed an independent judgement of greater significance when he appended to the paper an extension of some remarks on "Cohesion" he had made at a previous meeting of the Association. Henry usually fell short of eminence as a generalizer, but in this instance he displayed no small talent in the manner in which he assembled facts drawn from remotely associated experiments with liquids and solids. The value of this note does not reside in its deduction that tensile strength is of the same order in liquids and solids, or that the difference in rigidity is due to the molecules "slipping"; rather, attention should be focussed upon the inferred extent to which his conclusions were to be applied to Admiral Dahlgren's ironclads.

The note begins with a reference to the studies on the cohesion of liquids made at Princeton, and the application of the ideas he had derived from them to the breakdown of metals. These studies induced him to conclude that rigidity must be attributed, not to the commonly supposed attractive powers between molecules, but to their resistance to "slipping," or lateral cohesion. He contrasted the nature of the fracture of cast steel with that of lead. When the crystalline structure is perfect, as it nearly is in steel, the molecules are incapable of lateral movement; the substance is then rigid. It ruptures with a transverse fracture of the same size as that of the original section of the bar. He contrasted this with the pulling apart of a lead bar, where the molecules slip in lateral motion. The rod increases in length, and its thickness diminishes before the fracture takes place.

From this he deduced that "the form of the material ought to have some effect upon its tenacity, and also that the strength of the article should depend in some degree upon the process to which it has been subjected. . . . Metals should never be elongated by mere stretching, but in all cases by the process of wire-drawing or rolling." As a practical illustration of the truth of this deduction he cited the frequent breaking of locomotive axles, which was due to their shaping by hammering instead of by rolling.

The experiments from which these deductions were derived were performed in the United States Navy Yard, and Henry had the cooperation of John Adolph Dahlgren. The knowledge of metal processing was not lost upon the naval constructor.

Another paper read before the same association in the following year is a further example of the manner in which Henry applied his theoretical knowledge to practical purposes. This was published under the title "On Acoustics applied to Public Buildings." [6]

Henry and Bache had been examining the acoustical properties of the Smithsonian lecture hall, when the President requested them to extend their study to the plans Captain (later General) Meigs had prepared for additional rooms that were to be provided in the Capitol. The science of interior acoustics was entirely undeveloped, so that the investigators had to break new ground. Bache, Meigs, and Henry set off on a tour embracing all the principal halls and churches of Philadelphia, New York, and Boston. Each was submitted to inspection and tests.

The conclusions reached on this tour were submitted to experimental tests in which one detects the mind and hand of the indefatigable Henry. Much that was hitherto unknown about the propagation and decay of sound, its absorption, reflection, and reverberation, was brought to light.

It is inconceivable that the paper in which Henry presented the conclusions was allowed to languish in neglect, for he dealt with the subject in the manner of the modern acoustical engineer, but without benefit of the latter's instruments. A great deal of what he revealed had to be learned again seventy years later when the improved phonograph, radio, and sound motion picture gave to the science of acoustics a wider application. Instead of the microphone and electronic measuring devices, Henry had to employ a pair of clapped hands, a tuning fork, or his own voice. His ear was his measuring instrument. Yet the conclusions he reached are consistent with our modern conception of the satisfactory acoustical construction of a large room in which speeches are to be made and heard.

[6] A.A.A.S. *Proceedings.* Vol. X, p. 119; *S.W.* Vol. II, p. 403.

Incidentally, no one could be keener than Henry in devising and using measuring instruments. An illustration of his ability is given in another paper,[7] on the radiation of heat.

Taking a fifty-year-old paper by Rumford [8] which offered some conclusions without any account of the experiments from which they were derived, Henry designed and conducted his own experiments with measuring instruments based on Melloni's thermocouple. He was thus able to confirm by controlled experiment the conclusions which had been taken on trust for fifty years.

The final paragraph of his paper reads: "That a solid substance increases the radiation of a flame is an interesting fact in connection with the nature of the heat itself. It would seem to show that the vibrations of gross matter are necessary to give sufficient intensity of impulse to produce the phenomena of ordinary radiant heat. Also, since the light is much increased by the same process we would infer that by means of the solid the vibrations constituting heat are actually converted into those which produce the phenomena of light."

The pure scientist forces himself to the front in the suggestion that heat and light were but forms of radiant energy, but we are momentarily concerned with the practical scientist. Rumford had declared that the back and sides of the fire grate should be built of fire-brick, and never of iron if the heat of the fire is to be radiated. Henry gave this a practical test and confirmed Rumford's assertion. Was it entirely by chance that in the same year in which Henry revived the subject, Matthew Baird of the Baldwin Locomotive Works abandoned his idea of using a sheet iron deflector in a locomotive boiler and substituted one of fire-brick? [9] Science had neglected the steam engine up to this time, but the period had come when practical men, having exhausted their resources, were turning to the scientists for aid in improving the steam engine, and Henry let them know that the appeal was not in vain.

By far the most interesting example of Henry's uncommon blend of science and engineering is to be found in his work for the Light House Board. His interest in this field was undoubtedly stimulated by its association with the labors of Bache, who was also contributing to the safety of the mariner at the Coast Survey, where the reefs, shoals,

[7] *A.A.A.S. Proceedings.* Vol. IX, p. 355; *S.W.* Vol. I, p. 355.
[8] *Journal of the Royal Institution.* 1802. Vol. I, p. 28.
[9] *History of the Baldwin Locomotive Works, 1831–1923,* p. 57.

and rocks which menaced navigation were being charted. It became part of Henry's duties to see that these marine hazards were appropriately marked with lights.

He was appointed a member of the Board on its establishment by President Fillmore in 1852. Although he had many other duties to perform at the time, the mere fact that he was invited to participate in making coastwise navigation safer and surer was sufficient to enlist his active sympathy. He was placed in charge of all the experimental and test work concerning the service. This furnished him with highly diversified duties of no small responsibility to ensure that the signals, both by light and by sound, were of approved excellence and were attended by intelligent operators.

Before the establishment of the Light House Board, the national service had been conducted on a simple and inexpensive basis, its methods being modern only when compared with ancient fire beacons. The great extension of ocean-borne commerce required that the system be overhauled and modernized. At the time of Henry's death it equalled if it did not surpass the lighthouse services of Europe.

One of the first improvements recommended by Henry was the installation of scientifically designed Fresnel lenses in the lighthouses. This lens is constructed so that it refracts the light into parallel rays in the required direction. It permits a reduction in the weight of the glass used and showed a marked improvement upon the light furnished by an Argand lamp with a parabolic reflector. The suggested use of these admirable lenses might have been expected from the nature of Henry's physical studies, but his next employment on behalf of the Light House Board is of a very different nature.

The lighthouse lamps all burned sperm oil, which had advanced to a price which made it necessary to seek an economical substitute. Colza (or rape seed) oil had been introduced as an illuminant into European lighthouses, which were then regarded as superior to those in the American service. But the cultivation of rape seed in this country had not reached the stage where it could be depended upon to yield an unfailing supply of oil.

Lard oil had been suggested as a substitute for sperm oil because of its lower cost, but this was rejected when tests by an apparently competent authority, Professor J. H. Alexander of Baltimore, had shown it to be inferior as an illuminant.

Henry undertook to search for a suitable substitute. At the Light

House Depot on Staten Island he built a dark fire-proof chamber, painted black inside to ensure photometric accuracy for his tests. In this gloomy abode he passed the greater part of his vacations between 1852 and 1864 in testing various oils.

Measurements demonstrated that colza was about equal to sperm oil in illuminating qualities, but that lard oil was inferior. Most experimenters would have abandoned lard oil there and then. Henry merely laid it aside to experiment with the newly discovered kerosene. This he found to be too volatile, dangerous, and variable in quality. He rejected it, and resumed his investigation of lard oil.

Earlier in his career Henry had learned a good deal about capillarity. From his earlier studies he concluded that it was the absence of the ascensional attraction in the lampwick and its low fluidity which rendered lard oil a poor illuminant. He set himself to the task of improving fluidity and capillarity without impairing the illuminating quality.

Eventually he learned that under certain conditions of temperature the rejected lard oil was actually superior to sperm oil as an illuminant. The question arose: Could the requisite conditions be obtained in lighthouse lamps? A practical test demonstrated that the heat generated by the large Argand burners installed in the lighthouses rendered them ideal for the consumption of lard oil.

When this work had been carried to a successful conclusion, Henry turned his attention to another matter of grave concern to mariners. "Among the impediments to navigation none perhaps are more dreaded than those which arise from fogs," he wrote. "The only means at present known for obviating this difficulty, is that of employing powerful sounding instruments which may be heard at a sufficient distance through the fog, to give timely warning of impending danger."

Although the experiments made to obtain the best available fog signalling apparatus were not performed until after the end of the Civil War, they are dealt with here as further illustration of Henry's capacity as a practical engineer.

As everyone knows the descent of fog along the coast had rendered sound signals necessary wherever navigation was practicable. During the first half of the nineteenth century a miscellaneous assortment of instruments such as bells, whistles, horns, and guns had been employed to warn vessels of lurking danger. Many curious and often contradictory phenomena had been noted in the use of these instruments, especially in regard to variations of audibility under different conditions. Although the whole subject had received much attention among the other maritime nations, little or nothing had been attempted in this country. Therefore Henry's suggestion of a sound signalling system based on his observations and deductions were among the most valuable of his contributions to the betterment of the lighthouse service.

Having formed some conception of the nature of the problem, he began to experiment. To the traditional instruments of warning which he tested, he added steam whistles, a reed trumpet blown by a jet of air from a hot-air engine, and a steam operated siren with a revolving disc at the lower end of the horn. The experiments were too important to rely upon the ear for sound measurement. The human element had to be eliminated.

He elaborated upon the Sondhauss phonometer, or artificial ear, for making sound waves visible to the eye. The instrument as Henry used it comprised a sheet iron horn, with a thin membrane of goldbeater's skin stretched across the narrow end. Fine sand was strewn on this membrane, and when the received sound caused the sensitive membrane to vibrate, the sand was agitated. The agitation of the sand by the lower intensities of sound had to be observed through a lens.

He began this series of investigations at New Haven in 1865, where he observed that sound, moving against the wind and inaudible to the ear of a person on the deck of a vessel, became audible when the listener ascended to the mast-head. This was an observation that confirmed Stokes' theory of sound refraction due to wind.[10] This theory stated that upwind from the source of sound, the sound waves would be flattened and the sound "ray" deflected upward, so that the sounds would not be heard at distances as great as on the other side, where the effect is to make the wave more convex and to deflect the "ray" downward.

Henry's observation suggested that sound was more readily conveyed by the upper current of the air than by the lower. He drew some important practical conclusions from this principle of wind refraction, one being that a continuous sound, as from a horn or

[10] Stokes, G. G. Effect of Wind on Intensity of Sound. British Association *Report*. 1857, p. 22.

THE RIPENING YEARS

whistle, would be less likely to be lost by refraction due to adverse winds, than sounds of a short impulse, as from a bell or a gun. He also concluded that it was probable that sounds of high pitch would be more interfered with by refraction than those of medium or low pitch.

He explained at great length the refraction owing to inequality of the temperature of horizontal layers of air. This may be of the nature, he said, which will deflect sound waves either up or down, according as the upper or the lower stratum of air involved is at the higher temperature. Variations in the temperature of the atmosphere along vertical lines were shown to be sufficient to account for many acoustic phenomena with which many observers were familiar.

He established that beyond relatively limited distances sound reflection was ineffective, and he tested the various methods for projecting sound which induced him to place them in the order of the siren, trumpet, whistle, and bell. He therefore began to seek for the best form of siren.[11]

His experiments were all repeated by Professor John Tyndall in the interests of Trinity House, guardian of Britain's shores, who had to borrow some of Henry's apparatus, and who published the results of his experiments in a famous book on sound. Tyndall was particularly interested in the siren, which was Henry's improvement upon the design of Cagniard de la Tour. The English tests only served to confirm those made in America, and led to the adoption of the siren as the standard fog signal in Britain.

Henry obtained a certain amount of relaxation by using the vessels of the Light House Board to make exploratory voyages which permitted him to put his apparatus to test under actual operating conditions. While on these excursions he investigated the theories advanced by mariners and lighthouse keepers based upon empirical knowledge. It was in the course of these investigations that he detected the "belt of silence" or skip-distance at intermediate ranges, similar to that which occurs in the transmission of short wireless waves. He found Stokes' theory sufficient to account for the phenomena in sound, but he employed a number of free balloons to obtain experimental proofs.

[11] Much experimentation has been conducted since Henry's day and it now seems to be accepted that a sound resembling a powerful grunt or the bellowing of a mighty ox is best able to penetrate snow, hail, rain, or fog, and to arrest the attention of the mariner.

THE RIPENING YEARS

Arising from these studies was an observation which illustrates Henry's ability to grasp a piece of evidence and turn it to practical account. Efforts to study the velocity of the wind at higher altitudes had hitherto baffled observers. Henry had to find some method by which he could ascertain these velocities if he were to furnish the experimental proofs for Stokes' law. A swift inspiration gave him the means he needed. He timed the speed at which the shadow of a cloud sped across a piece of measured ground.

Henry's reports upon his experiments were all published in the annual reports of the Light House Board, and the more scientific aspects were chosen for expansion and explanation before the Washington Philosophical Society. On one of the occasions when Henry chose to address the members of this society, Tyndall was present as his guest. A mild difference of opinion arose between the two experts on sound over the influence of snow, rain, and fog upon the audibility of sound. Tyndall attributed many observed phenomena to "the existence of acoustic clouds, consisting of portions of the atmosphere in a flocculent, or mottled condition due to the unequal distribution of heat and moisture which, absorbing and reflecting the sound, produce an atmosphere of acoustic opacity."

While he cautiously refrained from denying that such atmospheric conditions might exist, Henry decided to hold by Stokes' theory, which he had been able to confirm by experiment and which, he thought, was sufficient to account for the phenomenon. This was a friendly divergence of opinion which might occur at any meeting of scientists, but it was unfortunately magnified by officious supporters of the two views, who contrived to irritate and exasperate each other in print almost to the stage of violence. Happily, their antipathy was not communicated to either of the two great physicists, who remained the best of friends, which was probably a good thing for Henry since, had he been disposed to deny Tyndall's theory, he would have encountered in the English scientist an adversary of such combative nature that he was rarely without a controversy upon his hands. Knowing the nature of the man, Henry very wisely did not press his own aspect of the case, but held his peace.

His services to the Light House Board were recognized in 1871, when Admiral Shubrick resigned from the chairmanship, and Henry was appointed his successor. Thereafter he regularly attended the weekly meetings of the Board in addition to devoting his summer vacations to its work in the laboratory on Staten Island or on the Board's vessels. In the summer of the year he was appointed chairman, when he was seventy-three years old, he made an inspection of the lighthouses on the California coast.

His long service to the Light House Board constituted by no means the least valuable portion of his labors in the interests of the nation. The high efficiency attained by the Light House establishments is due more to his intelligent administration and the high standards he devised, than to any other cause. The example he set in applying his wide knowledge of physical principles and utilizing it for practical ends has inspired the service to maintain an efficiency that is universally admired.

His researches on sound were a considerable contribution to that science. The more important conclusions he reached were summarized in 1878 in his last annual report to the Board. They were in effect:

The audibility of a sound at a distance depends upon the character of the sound; to secure audibility at a distance the transmitted sound should be of medium pitch; the loudness, depending upon the amplitude of vibration of the sounding body, should be as great as possible; and the volume of sound, depending on the magnitude of the vibrating surface, should also be great. Audibility also depends upon the state of the atmosphere, the most favorable condition being that of stillness and uniformity of density and temperature. The cause of deficiency or loss of audibility is the direct effect produced by the wind. While as a rule, the audibility of a sound is greater on the side toward which the wind blows, this is due to downward refraction, rather than to the simple carrying effect of the wind, which would be barely sensible. Those instances of greater audibility on the windward side are to be explained by reference to a dominant upper wind, opposite in direction to that near the earth's surface.

He rejected the idea that fog, snow, or rain materially interferes with the transmission of loud sounds. The siren he adopted was heard at a greater distance during a dense fog than during the prevalence of a clear atmosphere. Projecting portions of land or buildings may produce sound shadows, which accounts for some sounds becoming inaudible on closer approach. The existence of an "aërial echo" was established, probably due to the reflection of sound from the surface of the sea.

The outbreak of the Civil War found the federal government without any organized staff to handle the innumerable problems of a semi-scientific nature that arise in armed conflict after a period of peace which has witnessed many technical advances. At such a time a multitude of proposals have to be considered seriously that would be coldly ignored in more tranquil times.

Inevitably, the war produced a crop of inventors who deluged the President and the Secretaries of War and the Navy with suggestions for the improvement of weapons, methods of conducting military operations, the building of ironclads, and the means of navigation. The gulf between fancy and proof is often ignored by an ardent patriot with a suggestion for winning a war, but every idea submitted had to be scrutinized for fear that some valuable suggestion might be overlooked.

To relieve the harassed conductors of the war from the necessity of winnowing a little grain from the great volume of chaff, various boards of experts were set up. Henry and his technical staff were soon deeply involved.

Between the years 1860 and 1864 "several hundred reports, requiring many experiments and pertaining to proposals purporting to be of high national importance or relating to the quality of the multifarious articles in fulfillment of legal contracts, have been rendered. The opinions advanced in many of these reports not only cost much valuable time, but also involved heavy responsibilities," said Henry's summary of the work in his Report for 1864.

The loose arrangement of temporary boards to deal with the technological problems confronting the service departments soon proved inadequate, and interest was revived in suggestions which had often been made for the creation of a permanent national scientific organization composed of men of established reputation to whom the government might delegate its problems. The idea of a national academy of scientists had been in the minds of men like Washington, Jefferson, Franklin, John Adams, and others of scarcely less eminence, but it had been thrust into the background by Jacksonian democrats through the fear of strengthening federal government in the intellectual sphere. But the idea had never been entirely lost from sight by those whose vision was unclouded by political motives. When retiring from the presidency of the American Association for the Advancement of

Science in 1851, Bache had dwelt upon this need in expressing his belief that "an institution of science supplementary to existing ones is much needed in our country, to guide public action in reference to scientific matters." [12]

No action was taken. But the crisis created by the outbreak of the war between the states found the federal government so ill-prepared that there was wide-spread regret that Bache's suggestion had not been heeded in official quarters.

One man who, as a member of various temporary boards, moved among the groups of technical men, was Commodore (later Admiral) Charles H. Davis, then serving as chief of the Bureau of Navigation, Navy Department. As a lieutenant he had attended the Albany meeting at which Bache advocated the creation of a national scientific society. Now that a crisis had rendered its need more fully and practically apparent, Davis thought he saw a way of bringing into existence an organization of the required character. He discussed the matter with Henry and Bache, but neither thought the time was propitious, when the country was plunged into war and its leaders were fully occupied.

As chief of a government bureau compelled to enlist the aid of scientists, Davis was anxious to see some system introduced to replace the disorder then prevailing, and finding that his proposal of a national academy did not meet with the approval of his two scientific acquaintances, he suggested an arrangement similar to the Select Committees appointed by the British parliament in dealing with such emergencies.

Although both men would have preferred a national academy, Henry and Bache agreed that a small Permanent Commission, to which the government's problems of a scientific nature might be referred, had a much brighter prospect of being accepted.

The necessity of such a permanent commission may be gathered from the speed with which it was set up after it was suggested.

The idea of the commission seems to have occurred to Davis shortly before February 2, 1863, on which day he records in his diary that Henry, Bache, and he had conferred upon it.[13] Henry was so im-

[12] A.A.A.S. *Proceedings*, 1851, p. xlvii.
[13] The version of the affair given in Davis, C. H., *Charles Henry Davis: Rear Admiral*, 1899, should be supplemented and corrected with that given in True, F. W., ed. *History of the first Half Century of the National Academy of Sciences*. Washington. 1913, p. 4ff.

pressed with the idea that he promptly submitted it to Gideon Welles, Secretary of the Navy, who likewise perceived its merits. On February 11, he issued a letter of appointment to Henry, Bache, and Davis as members of the Permanent Commission. The letter, preserved in the archives of the Navy Department, reads:

> Navy Department
> Washington
> February 11, 1863
>
> Sir:
>
> The Department proposes to organize upon the following program a permanent commission to which all subjects of a scientific character on which the government may require information may be referred.
>
> Propositions relative to a permanent scientific commission:
>
> 1st. There shall be constituted a permanent commission consisting of, for the present, Commodore Davis, Professor Henry, and Professor Bache, to which shall be referred questions of science and art upon which the Department may require information.
>
> 2nd. The commission shall have authority to call in associates to aid in their investigations and inquiries.
>
> 3rd. The members and associates of the Commission shall receive no compensation for their services.
>
> You are directed to act as a member of the Commission in conjunction with Commodore Davis and Professor Bache.
>
> Such matters as are presented to the Department will be referred to you for examination and report by the Commission.
>
> I am respectfully, etc.,
> Gideon Welles,
> Sec'y of Navy.[14]

The Commission was of very short duration. Eight days after it had been appointed, the National Academy of Sciences was unexpectedly formed without Henry's knowledge.[15] He was, of course, chosen as one of the fifty original members.

[14] Letters to Heads of Bureaus (manuscript). Vol. IV, p. 153.
[15] The circumstances are treated in the next chapter, which deals with Henry as an organizer of science.

The members selected were scattered over a wide area of territory, and it was obvious that prompt and effective action could be undertaken only through the medium of committees of specialists. Henry was appointed chairman of the first committee elected at the solicitation of the Secretary of the Treasury to consider "The Uniformity of weights, measures, and coins, considered in relation to domestic and international commerce."

This committee was not discharged until 1866 (a standing committee was appointed in the following year) and, although their proceedings were not fully reported, Bache gave an indication of the course of the deliberations when he, as president of the National Academy, reported: "The discussions were strongly in favor of the adoption of the French metrical system, but more strongly, in fact unanimously, in favor of the effort to arrive at a thorough international system—a universal system of weights, measures, and coins, available for the general acceptance of all nations." [16]

In view of the unsatisfactory international situation it is a tribute to the detachment of the members of this committee that they could separate themselves from the cross currents of the conflict at home with all its implications abroad, and think of a time when all nations would be at peace and in search of ties that would bind them together.

The committee's first report was transmitted to the Secretary of the Treasury, Hugh McCulloch, on February 17, 1866, accompanied by a letter from Henry in which he said:

> . . . The subject is one of much perplexity. While, on the one hand, it is evident that a reform of our present system of weights and measures is exceedingly desirable, on the other, the difficulty of adopting the best system and of introducing it in opposition to the prejudice and usages of the people is also apparent.
>
> The entire adoption of the French metrical system involved the necessity of discarding our present standard of weights and measures—the foot, the pound, the bushel, the gallon—and the introduction in their place of standards of unfamiliar magnitudes and names.
>
> Such a change, in my opinion, can only be, in a government like ours, the work of time and through the education of the rising

[16] Nat. Acad. Sci. *Report for 1863*, p. 4.

generation, for this purpose, should the Resolution now before Congress to establish a bureau of education be adopted, the French metrical system might be taught under the sanction of the government in all the common schools of the country.

The system, however, is not considered by many as well adapted to the Anglo-Saxon mind as one which might be devised, and it was therefore the opinion of a minority of the Academy, that, could England and the United States agree upon a system for adoption, it would in all probability in time become universal.

The agreement in favor of the French metrical system is, however, that it has been already adopted in whole or in part in several nations.

The report and the Academy's recommendations, together with a summary of the history of weights and measures in European countries was presented to the House Committee on the subject and some compromise bills were passed by Congress.

Henry also served on the third committee appointed by the Academy, which was set up May 20, 1863, at the request of the Navy Department "to make experiments on the local attraction experienced in vessels built wholly or partly of iron."

Iron ships were fast replacing those built of wood, and many of the eighty-eight vessels on the naval list had wooden hulls with iron plate protection above the waterline, iron decks, or iron parts in the superstructure. The presence of these iron masses was the cause of large and variable compass deviations which made navigation erratic and, at times, precarious. Strange and amateurish efforts had been made to overcome this difficulty, such as installing several compasses which were sometimes mounted in pairs, setting the compasses in iron pots four inches thick, placing them in zinc tanks packed with charcoal, and similar tricks of equal fantasy.

Perceiving the need for haste in presenting their recommendations, the committee realized that a method adopted in Britain at the suggestion of Robert Airey, the astronomer, was the best then available. This consisted of counteracting the local attraction by means of bar magnets placed in suitable locations. This was as far as the committee could go, although it must be said they spent no time upon a leisurely investigation of their problem. It is strange, however, that the problem which proved too much for the combined minds of Henry, Bache,

Wolcott Gibbs, Benjamin Pierce, and W. P. Trowbridge was solved single-handed by Sir William Thomson, who shortly afterwards produced the improved mariner's compass which speedily became standard equipment for ships of every country.

Nevertheless, the committee spent a busy summer inspecting and correcting the compasses on twenty-seven naval vessels distributed between Boston, New York, Philadelphia, and Hampton Roads. They set up a compass adjusting station at the Philadelphia Navy Yard. A detailed report of their labors covered seventy-three printed pages.

On April 8, 1864, at the request of the Secretary of the Treasury, the Academy appointed a committee "to examine and report upon aluminum, bronze, and other materials for the manufacture of cent coins." Henry was an active member of this committee. The purpose was really quite different from the terms of reference. There was no intention of replacing the millions of one and two cent pieces in circulation. The committee was expected to determine the properties of aluminum bronzes with a view to employing them in making coins of other denominations. The experiments carried out by the members of the committee were suggested by a claim advanced in France that aluminum added to silver would prevent the latter from tarnishing and would give it greater hardness.

This committee was not called upon to perform any arduous investigation. Henry undertook to do the research, which proved to be exceedingly simple. He had a bar of aluminum bronze made by an old friend of Princeton days, Joseph Saxton. This was conveyed to the mint where it was converted into coins. To determine the relative hardness of the metal, he had some coins made of the aluminum bronze alloy, some of pure copper, and some of legal bronze. These were all thrown into boxes and shaken together for a long time. The new alloy emerged from this test with credit, but it failed dismally in the tarnishing test. Held for some time in "a sweaty hand," it acquired an ugly discoloration.

Metallurgy was still in such a primitive stage that these practical tests were the best that could be devised at the time.

A task involving more laborious investigation was imposed upon Henry when the Secretary of the Treasury asked the Academy to appoint a committee to report upon the best method of proving and gauging alcoholic liquors, so that rules and regulations might be established to ensure a uniform system of inspection of spirits liable to taxation.

This was a delicate problem, even for a scientist without prejudices.

Excise and internal revenue taxes were unpopular in the United States because they were reminiscent of the despotism of colonial administration. The imposition of such a tax had led to the Whiskey Rebellion of 1794. This tax had been abolished by Jefferson, but was reimposed to meet the expenses of the War of 1812, after which it was again discarded. The vast expenditures during the Civil War necessitated a revival of the unpopular tax.

The collection of a high revenue tax on domestic and imported spirits called for much vigilance, an improved form of inspection, and better proving instruments. The Tralles hydrometer, then in use, gave percentages in alcohol instead of in "proof spirit," or in a mixture of half alcohol and half water, upon which commercial transactions were based. This form of hydrometer was not accurate, nor was it suitable for making readings under some conditions that prevailed. Moreover, the Treasury tables used for computation were suspected of inaccuracy.[17] The tables were intended to lighten the inspectors' labors in making the necessary conversions. The committee, of which Henry was chairman, was given the dual task of finding a superior hydrometer and of compiling new tables.

Appointed February 15, 1866, the members presented their report on July 21. In the annual report of the Academy [18] it was said regarding the work of this group:

> The duty devolved upon the members of the committee was one of much labor and responsibility. The tables accompanying the report are of much value and will be referred to by all persons engaged in pursuits requiring a knowledge of specific gravity and volume, at various temperatures, of alcoholic spirits of different strengths; they are not only indispensable to the distiller, rectifier, and gauger, but will prove extremely useful in the laboratory of the chemist, and in many processes of manufacture involving the use of alcohol.

[17] True. *Op. cit.*, p. 240.
[18] For 1866, p. 3.

Although this disposed of the work relative to the proving and gauging of spirits, the Inland Revenue Commission was still vexed with the problem of fraud, practiced either separately by the distillers or in collusion with inspectors. There was a suspicion that fraud was practiced on such a widespread scale that the government was being deprived of a large sum in revenue. The committee, therefore, was asked to suggest plans for outwitting the guile of the distillers and for frustrating the corruption of the inspectors.

Between June 1866 and January 1868 the committee examined plans and models of eighteen meters designed to be affixed to stills for measuring the output. All of this work fell upon Henry and E. Hilgard, the Acting Superintendent of the Coast Survey, as their associates had all withdrawn from active work on account of ill-health. As a result of their tests a meter was recommended for adoption, and a confidential statement was issued on the measures to be adopted for defeating the attempts at manipulating these meters.

No sooner had the selected meter been put into operation than a storm of opposition arose from the distillers. Congress hastily appointed its own committee to serve with Henry and Hilgard in examining all meters presented to them, and to make recommendations in the best interests of the government. The introduction of the resolution to appoint this committee was the occasion for an unusually acrimonious debate in both Houses of Congress, during which the integrity of the highest government officials and the qualifications of scientists as practical men were questioned in the frankest terms.

The new committee approved of the meter recommended by Henry and Hilgard, but the distillers did not abate their opposition to the use of any meter, and after five years of ceaseless agitation its use was abandoned. The reason was never divulged, but it may be surmised.

No one appears to have given serious attention to Henry's right to be regarded as one of the pioneers of aeronautics in this country, yet it was largely through his influence and encouragement that the first aeronautical service was organized.

Henry had been introduced to the science of ballooning by John Wise, who had frequently consulted him with regard to meteorological problems confronting the balloonist. They had established such cordial relations that Wise named one of his balloons "The Smithsonian" and in 1859 used it to make ascents for the purpose of gathering meteorological observations at the suggestion of Henry.

As a result of his interest in Wise's projects, Henry was introduced to T. S. C. Lowe. This courageous aeronaut had conceived the idea of making a flight across the Atlantic, encouraged by the belief that an easterly air current existed which warranted assuming the risk of what the uninitiated looked upon as a foolhardy enterprise. Other balloonists, not excluding Wise, scoffed at the proposal as being unworthy of serious consideration.

Lowe refused to be diverted from his plan. Determined to carry out the enterprise if he could raise the necessary funds, he approached a group of influential Philadelphians to gain their financial support. This group addressed a testimonial on his behalf to Henry, expressing their confidence in Lowe and urging that the Smithsonian furnish him with advice and aid. When they received this plea, the Regents gave it serious consideration, but instructed Henry to inform the group that the Smithsonian endowment did not permit the appropriation of funds for the purpose, but that their secretary would be ready to furnish Lowe with any assistance he required in the form of technical advice or service.

Henry wrote to Lowe approving of the latter's proposal to take advantage of the upper easterly air current for his projected voyage, and did not hesitate to express his belief that this might be used to make the passage of the ocean.[19] However, his characteristic caution moved him to recommend the aeronaut to make a trial voyage overland before attempting the more hazardous ocean journey, the success of which was wholly dependent upon the steadiness of a current of air which had never been fully tested. Henry thought the balloonist should acquire some experience of navigation in the current before undertaking a protracted voyage on which an accident was almost certain to be fatal.

Lowe accepted this letter as a scientific endorsement of his enterprise. He immediately dashed off to Washington for a conference with Henry, who was so impressed with the balloonist's knowledge and experience that he became an ardent supporter and a constant adviser. Together they laid plans for a trial flight which would furnish tests for the velocity, direction, and constancy of the easterly current.

[19] *Smithsonian Report for 1861*, p. 118.

This flight was made from Cincinnati but, unfortunately, Lowe was carried behind the Confederate lines to Columbia, S.C., where he was arrested as a spy. Nothing but the prompt intervention of someone who had learned of his association with Henry saved him from being tarred and feathered by those members of the populace who did not read the scraps of scientific information published in the southern newspapers.

However, the flight was judged by Henry to have provided convincing evidence that the balloon had valuable uses. He wrote a letter to Lowe after the latter's release suggesting that the balloon would be of advantage to the government in assisting their reconnaisances of the country around Washington.[20] Thus, we have evidence that the first suggestion for utilizing military balloons in the war originated with Joseph Henry.

In the subsequent proceedings, when Lowe hastened to follow Henry's suggestion to offer his services to the government, Henry was to learn, if he did not already know, the singular course of conduct followed by Lincoln's Secretary of War, Cameron.

Before approaching the authorities, Lowe outlined a scheme for a military balloon service which he submitted to Henry for approval. Henry examined it and gave his cordial approbation. He was so anxious that Lowe's proposals should receive serious consideration that he asked his friend Chase, the Secretary of the Treasury, to transmit them to Cameron. Chase discharged his mission with such expedition that, on the following day, Cameron sent for Henry, invited his opinion upon Lowe's experience and equipment, and requested him to attend and report upon a trial ascent Lowe had volunteered to make.

In his anxiety to secure the highest support for the scheme, Henry accompanied Lowe on a visit to the President, who listened attentively to the proposals of his two visitors, and promised he would give the project his support.

This experimental flight was made on June 1, 1861. Still determined to give Thaddeus Lowe all the support organized science could provide, Henry brought along the entire Smithsonian staff to give the aeronaut encouragement. This was the occasion on which the tele-

[20] Haydon, F. S. *Aeronautics in the Union and Confederate Armies*. Baltimore, 1941. Vol. I, p. 169. This valuable work gives ample quotations from Lowe's unpublished papers.

graph was used to transmit from an aerial station for the first time, and Lowe sent his famous message to the President, who was among the spectators, containing the words: "The city with its girdle of encampments presents a superb scene."

Thirty years after he had constructed the first rude electric telegraph found Henry assisting at the development of another science which was not to attain its full significance until another and even greater war called upon the scientist's resources. He stood midway between the generations, and he was as enthusiastic over the triumphs of the new as he had been over those of the old.

Henry submitted a detailed report upon the ascent to Secretary Cameron, as he had been requested to do, declaring that the balloon was so well constructed that, when filled with illuminating gas, it would retain its lifting power for several days. He pointed to the high mobility of Lowe's balloon, since it could be hauled by a detail of men over roads to and from its gas supply. In a third point he dealt with the range of observation, which extended over a radius of twenty miles with good visibility. Attention was also drawn to the possibility of making free flights at lower altitudes over the enemy lines, and then of rising into the upper easterly current to make a return flight.

If a modern reader with military experience should imagine that Henry had overlooked the balloon's utility for artillery "spotting," he may be reminded that the very limited ranges of Henry's day enabled the gunner to observe the effects of his own fire.

Henry's report was a complete approval of both Lowe and his equipment. Yet, in that inexplicable way of war departments, when the army gave its order for a balloon it was given, with Secretary Cameron's consent, to John Wise.

Henry was fully aware of Wise's experience, but he had thrown his support to Lowe because he perceived that his equipment was superior to that of Wise for military purposes. Wise had made some progress with the Topographical Bureau in his efforts to persuade the military authorities to adopt the balloon for intelligence work. When the Bureau's chief, Major Hartman Bache, expressed his interest and sought unprejudiced references upon the quality of the equipment offered, Wise did not hesitate to offer Henry's name as a referee. Since this was acceptable as the highest scientific recommendation he could have, the chief of the Topographical Bureau did not bother to consult with Henry. He accepted Wise's word, bought the balloon, and engaged the services of the aeronaut. Had he taken the trouble to refer the matter to Henry he would have learned that he could have struck a better bargain.

The disappointed but not discouraged Lowe then moved his balloon to the Smithsonian grounds, where he proceeded to submit it to additional tests under the critical eyes of Henry and his staff. Presently another aeronaut, Helm of Providence, who had been recommended to the government, was sent with his balloon to join Lowe, and Henry was requested to report upon the newcomer. Helm did not make such an impression upon Henry that he withdrew his support from Lowe, while keeping a kindly eye upon Wise's achievements, or rather his repeated failures.

For Wise was a failure. The fault cannot be laid to either a deficiency of the man or a fault of his equipment. Stupid orders by senior officers who chose to disregard his protests that the weather conditions were entirely unfavorable for balloon ascents, resulted in the destruction of all his balloons. The authorities were so discouraged that they were ready to abandon the incipient air arm.

At this critical moment Henry, who had first suggested the employment of balloons for military purposes in the Civil War, decided to intervene. He wrote a letter to the Secretary of War protesting against the abandonment of the balloon by the army, and urged that further trials with Lowe's improved equipment be conducted.[21] Henry's name was too potent to be ignored when attached to a recommendation of a scientific nature. The letter was passed down to Major Bache, who was in charge of balloon operations. Fortunately, he retained some faith in the balloon, despite the failures of Wise.

Bache was uncle to Henry's old friend Alexander Dallas Bache and he held Henry's judgement in high esteem, having worked with him for the Light House Board. He thereon revived the whole question, learned that Wise had been ordered by a few individual generals, McClelland among them, to make ascents and had obeyed with praiseworthy disinterest but with calamitous results to his equipment. Bache was favorably impressed by the reports of other officers who held more reasonable views upon the balloon's limitations. He made arrangements whereby Lowe's irregular relations with the authorities were placed upon a more satisfactory foundation. He also

[21] Haydon. *Op. cit.,* p. 197.

ordered balloons to be made upon Lowe's specifications and under his direction. Thereafter the history of the army's first aeronautical service belongs to Lowe's story, but it would be doing an injustice to Henry if we omitted to take into consideration the advice and aid he gave the aeronaut, and the influence he wielded when he chose to intervene at the critical moment in order to rescue the balloon service. Lowe certainly never withheld any credit from Henry for his share in promoting and maintaining the idea that the balloon had a place in military tactics.

ORGANIZER OF SCIENCE

BECAUSE the nature of scientific work demands a continuous exchange of ideas, scientists have always tended to form societies, formal or informal. Even the ancient scientist-philosophers gathered their little groups or "schools" around them, and the era which marks the rise of modern science also saw the beginnings of the great scientific societies which are still active today. In England the Royal Society was formed in 1662, and later the Royal Institution and the British Association for the Advancement of Science. In France the chief scientific organization was the Academy of Sciences. In the New World the first great American scientist, Benjamin Franklin, in 1727 formed an informal "club of mutual improvement" called the Junto, which became the American Philosophical Society in 1743. Bostonians could not long endure the preeminence of Philadelphian science, and in 1780 the American Academy of Arts and Sciences was founded in "the Hub." John Winthrop, Hollis Professor of Mathematics and Natural Philosophy at Harvard, and John Adams were the leaders in its organization. Most of the budding crop of scientific societies were stimulated by local pride as well as interest in science, and only gradually did some of them become national in character. We have seen that in Henry's early days even the little city of Albany had two scientific societies, and by the time Henry had attained eminence at the Smithsonian there were scores of scientific societies, big and little, scattered about the country. We were a nation of joiners, and societies were born and died almost too rapidly to be counted. Henry played an important part in nationalizing and internationalizing American scientific activity, both directly through his scientific work and through the influence of the three national and international institutions which he helped to organize—the American Association for the Advancement of Science, the National Academy of Sciences, and the Smithsonian Institution. The details of the founding of the Smithsonian have already been described at length, and further discussion in this chapter will be brief; but Henry's other activities in organizing and stimulating science must be considered in detail, for his contribution in this sphere, though difficult to evaluate precisely, is perhaps as important as his individual scientific work.

American science in the first half of the nineteenth century offers the names of several men who had made substantial contributions in various fields. Silliman, collecting minerals and establishing a chemical laboratory at Yale, while publishing the principal American scientific journal; Audubon wandering in the wilds to establish an international reputation as an ornithologist; Agassiz at Harvard laying a firm foundation for study and research in zoology; Maury exploring the mysteries of hydrography; Asa Gray, prominent in the study of flora. These men, save Audubon, were Henry's friends and, like him, had eventually been recognized and honored in Europe.

The natural sciences were flourishing. The efforts of Americans to employ chemicals to alleviate pain cannot be overlooked. The almost tempestuous developments in physics, in which Henry's own contributions had played no inconspicuous part, involved all the other sciences in their onrush. At all points one perceives incisive progress and a breaking away from older views.

If the names of P. T. Barnum and Tom Thumb were better known to the general public in this period than those of the men we have just mentioned, it is not to be implied that the nation was oblivious to the scientific effort being exerted. Between 1812 and the outbreak of the Civil War, numerous clubs, societies, lyceums, and associations for the study of science, philosophy, and the arts had multiplied. Henry favored the founding of these societies if "the object should be kept constantly in view, and care taken that it be not interfered with by a tendency to waste the time of the meetings in the discussion of irrelevant matters. . . . There is a tendency in this country to express little thought in many words, to cultivate a talent for debate, or the art of making the worse appear the better course—which is by no means favorable to either the increase or the diffusion of knowledge." [1]

---

[1] Extract from a letter to a correspondent asking advice on the organization of a local scientific society. *Smithsonian Report for 1875*, p. 217.

He went on to stress that the object of these local societies should be to make "the special knowledge of *each* the knowledge of *all.*" The farmer should be encouraged to talk about his crops and soil, the mechanic about his processes with a view to educating every member to exercise his powers of observation. "It is of vast importance to an individual that he be awakened to the consciousness of living in a universe of most interesting phenomena, and that the one very great difference between individuals is that of *eyes* and *no eyes.*"

But this related to the diffusion of knowledge only, and Henry would have the small local societies play their part in the increasing of knowledge by collecting information on the flora, fauna, geology, meteorology, and history of the region in which the society was located.

> Man is a sympathetic being, and no incentive to mental exertion is more powerful than that which springs from a desire for the approbation of his fellow men; besides this, frequent interchange of ideas and appreciative encouragement are almost essential to the successful prosecution of labors requiring profound thought and continued mental exertion. Hence it is important that those engaged in similar pursuits should have opportunities for frequent meetings at stated periods. This is more particularly the case with the cultivators of abstract science who find comparatively few fully capable of appreciating the value of their labors, even in a community how much soever enlightened it may be in general subjects.
>
> The students of history, of literature, of politics, and of art find everywhere men who can enter in some degree into their pursuits, and who can appreciate their merits and derive pleasure from their writings and conversation; while the mathematician, the astronomer, the physicist, the chemist, the biologist, and the student of descriptive natural history, meet with relatively few who can sympathize with them in their pursuits. . . . To them the world consists of a few individuals to whom they are to look for that critical judgement of their merits which is to be finally adopted by the general public, and with these it is of the first importance that they should have more frequent intercourse than that which arises from casual meetings.[2]

[2] On the Organization of a Scientific Society." Wash. Phil. Soc. *Bulletin.* Vol. I, p. vi.

Henry was speaking of the formation of small local societies. They were the tiny rivulets which flowed into the larger tributaries that united to feed the great river of national ideas. They were small and weak, but they were encouraged at the Smithsonian to keep alive the spirit of inquiry. The means of consolidating the gains and laying the lines of the onward march were still wanting, but Henry labored to provide the means, partly through the Smithsonian and partly through the organization of national societies of specialists.

The federal government's attitude toward the encouragement of science was that of strict utilitarianism. It established a precedent in 1832 when it appropriated funds to assist in an investigation conducted by The Franklin Institute into the causes of boiler explosions, but it had not continued to support this form of research. While not averse to finding funds for surveys and expeditions, the government was niggardly in making the reports of the explorers and surveyors available to students. Even the authors of the reports had cause to complain about the distribution. James Dwight Dana, who worked for four years with the Wilkes Expedition in the South Seas, did not receive a printed copy of the report he had labored to prepare.

Established by government, provided with a Board of Regents drawn from the most eminent men in the country, it might easily have been possible for Henry to make the Smithsonian into an exclusive agency of limited membership claiming precedence over other scientific organizations. Nothing of the kind was attempted. On the contrary, a policy was followed of maintaining friendly relations with any agency devoted to the encouragement of knowledge. Without the power to distribute patronage or to bestow awards, without the disposition to criticize or correct, the Smithsonian became a powerful auxiliary to scientific culture.

The "Plan of Organization" drawn up by Henry was a masterly exposition of the steps necessary to cope with the existing situation, and the manner in which he began to discharge his administrative duties gave assurance to his sincerity. It is difficult in these days of easy communication to appreciate the isolation of the rural districts, and now that handsome endowments have done so much to promote higher education, it is not easy to imagine conditions when few men had thought to endow it. The Smithsonian's slender resources were employed to the utmost to fill the abyss of neglect. As libraries and museums were founded, the Institution made gifts of books and loans of specimens. Local societies were encouraged, as we have observed, to conduct investigations into natural history and archeology; and equally important, they were encouraged to print the reports of their proceedings and to avail themselves of the Institution's system of exchanges. In this way the name and the work of the Smithsonian was brought more intimately to notice than any other organization, and the local societies were made to feel that they were a part of the worldwide promotion of science through this great institution.

During the first half of the nineteenth century science was beginning to harvest the fruits of one hundred and fifty years of growth. Henry saw that as basic science was carried over into applications, applied science would grow, and this would provide a further impetus for basic research. Already the activities of professional societies, laboratories, schools, colleges, hospitals, and observatories had created a complex situation. Henry knew that complexity would grow, and it was necessary to provide central integrating forces.

Henry's merit lay in perceiving the means of meeting this situation. He knew that there would be a vast amount of duplicated effort unless there were free and accessible channels for interchanging ideas.

The formation of technical societies holding regular meetings would help to bring men interested in the same subjects together for the interchange of ideas, for discussion of technical problems, for reporting on work in hand and projected so that duplication might be avoided and new avenues opened, and for taking whatever group action seemed advantageous.

Even before Henry went to Washington to direct the activities of the Smithsonian he had begun to play a part in the movement for integration. There were numerous societies devoted to general studies but only tentative efforts had been made toward the formation of national bodies. A few specialists, as the geologists and geographers, had formed national groups, and it was from one of these that the first movement toward the creation of a great national society of scientists had its origin.

As early as 1838 Henry had written to Dr. J. C. Warren approving the suggestion that scientific workers in this country should form a society corresponding to the Association for the Advancement of Science which had been formed in Britain. The two men were somewhat in advance of their times, but when the opportunity to take advantage of a favorable situation appeared, both were at hand to embrace it.

Henry joined the Association of American Geologists as an associate member when it was established in 1840. He attended no meetings until the third, in 1842, when it was decided to change the name to the American Association of Geologists and Naturalists, as indicative of the members' desire to extend the field of their activities. This was still not broad enough to satisfy Henry who wished to see the Association widen its activities to cover fields in technology.

He was active in the arrangements which were made by Warren to extend the group until it bore a resemblance to the well established British Association for the Advancement of Science, whose meeting he had attended in 1837. He attended the meeting of the Geologists in 1847 where it was decided to change the name to the American Association for the Advancement of Science, which would permit membership for all those engaged in a wider variety of sciences. From that time onward he was regular in his attendance, took a lively interest in the organization and its proceedings, and contributed a number of papers relating to the research he conducted for government departments.

The high position he had gained in the opinions of his fellow workers singled him out for posts of honor. He presided over the meetings of the physical section in 1848, and in the following year was elected to the presidency of the Association. At one time he held the curious distinction of being a committee of one elected to rule on "scientific ethics." This put him in the position of being something of a dictator but, unfortunately for scientific morals, he never issued any report. Nevertheless, the appointment serves to show in what high esteem his conduct was held by his contemporaries, who trusted him, confided in him, watched his actions, and were willing to follow his example.

His position as a leader and the manner in which he exercised his influence in bringing about and encouraging the organization of groups is unique. His official position at the Smithsonian brought him into contact with most of the men in the United States who were engaged in scientific studies, and he established personal relations with those men at the meetings of the learned societies. The inspiration he provided and the control he exercised over these bodies was constant and far reaching. It has never been equalled since his day by any single individual.

The significance of the duties he undertook as a leader cannot be appreciated fully until it is recognized that at the outset the professional scientist was a rarety; the public was barely conscious of his existence. The scientist was usually an amateur. He had followed some other profession, such as medicine or teaching, and the study of science had been an avocation rather than a regular vocation.

By the time of Henry's death the men who were engaged in the fields of science were organized in articulate groups, showing to the lay world that their numbers had increased, that the scientists of the country were no longer a few stray individuals, but were a strong body with discipline and traditions. When a scientist spoke of his work at a meeting of his colleagues, his was no longer a solitary voice crying in the wilderness. He was surrounded by a crowd of witnesses and, if his researches met with their approval, his opinions thereby gained greatly in weight.

Apart from this, the individual gained because these associations kept him alert to developments in all science. There was a variety in the papers read which is missing from similar meetings today because the societies were then unaffected by the withdrawal of specialists to form their own organizations. A man's knowledge was not so intense, it is true, but it was broader, and interest was sustained in the wider outlook by injections administered by men outside their own fields.

These advantages, so apparent in retrospect, were not so obvious when Henry threw himself into the task of organizing science. The value of his contribution toward this end has been summarized by the astronomer Edward S. Holden, who says:

> [The] enumeration of specific benefits does not convey an adequate idea of the immense influence exercised by the [Smithsonian] Institution upon the scientific ideals of the country. It was of the first importance that the beginnings of independent investigations among Americans should be directed toward right ends by high and unselfish aims. In the formation of a scientific and, as it were, a moral standard a few names will ever be remembered among us, and no one will stand higher than that of Henry. His wise, broad, and generous policy, and his high personal ideals were of immense service to his colleagues and to the country.[3]

[3] "Beginnings of American Astronomy." *Science* (n.s.). 1897. Vol. V, p. 933.

ORGANIZER OF SCIENCE

Henry was not satisfied with the organization of scientists among themselves, he strove to raise their status in the eyes of the public. This aspect of his work came to light during his presidency of the National Academy of Sciences. Henry did not play any conspicuous part in the establishment of the Academy. In fact, one might suppose that he had reason to feel aggrieved at the conduct of his intimate friends in this episode.

Reference has already been made to the formation in 1863 of the Permanent Commission to advise the Navy Department (p. 258). Both Henry and Bache would have preferred to see an academy composed of the best minds in the country, such as Bache had proposed. But even after the discussion of the proposal with Commodore Davis, the two friends were of opinion that the Commission had a more reasonable chance of acceptance during the critical days of war. Henry must have felt this strongly, for it was he who carried the suggestion to create the Commission to the Secretary of the Navy.

Eight days later, the National Academy of Sciences was formed without Henry's knowledge.

The circumstances which brought about the sudden change in plans center around the arrival in Washington of Louis Agassiz, who had come to attend a meeting of the Smithsonian Regents, to which he had been appointed. At first blush it would appear that the occasion was marked by a treacherous act committed by Bache and Agassiz against their mutual friend. The situation is outlined in a letter from Torrey to Asa Gray.[4]

Agassiz was to have been the guest of the Henrys during his visit to the capital but for some reason or other he went directly to Bache's home, where he remained to dine. The only other guest known to dine with Bache that day was Senator Henry Wilson, of Massachusetts. The conversation turned upon the suggestion of forming a new academy upon the outline proposed by Commodore Davis and, before the dinner guests dispersed, they had enlarged upon that original plan.

On the following day Senator Wilson gave notice in the Senate and received approval to introduce a bill to incorporate the National Academy of Sciences. The bill was signed by the President on March 3, 1863. Henry did not learn of it until two days later. Although Gray

[4] Rodgers, *Torrey*, p. 274.

and Torrey, two of the charter members of the Academy, were suspicious about the meeting at Bache's home, from which their friend Henry had been absent, the incident did not in any way affect the cordial relations between Bache, Agassiz, and Henry.

Henry was fully convinced that the formation of the Academy was desirable. Why, then, did his name not figure among those of the prime movers of the project? Not because he entertained any objection to the motives, nor to the form in which the proposals were advanced. Of this we can be sure; but Henry was a realist who did not shed his realism under pressure. They were living in a time of crisis, and he must have told Bache and Davis in unmistakable terms that he did not believe the time was ripe to worry Congress with a proposal to create the Academy. Once Henry had made up his mind upon a matter he was exceedingly hard to dislodge.

Although in a letter dated March 9, 1863, addressed to Stephen Alexander, he expressed dissatisfaction with the list of members, yet from the moment of its inception Henry was a staunch supporter of the Academy. He was on the program to read a paper at the first meeting, held in the Capitol in 1864, and he was actively engaged in discussing with Wolcott Gibbs how the proceedings of the Academy should be printed and published. He was disinclined to employ the meager funds of the Smithsonian for the purpose, except with regard to appropriate special memoirs which the Academy might wish to have published. Even these would have to bear the Smithsonian imprint, and Henry did not hesitate to inform Gibbs he regarded this as unworthy of the Academy.[5]

Bache was the first president of the Academy and James Dwight Dana vice president. As both were in ill-health, the duty of writing the annual report for 1865 fell upon Henry. Illness continued to handicap Bache, and Dana resigned, so that Henry, in spite of pleas that he was much too busy to attend to Academy business, was elected to the vice-presidency and charged with performing the duties of the presiding officer. In view of his long occupation of the presidency of the Academy, it is amusing to find him on this occasion pleading to be released from the office as soon as Bache was restored to health. Bache did not recover. On his death in 1867, Henry was elected president, an office he retained until his death eleven years later.

His first annual report, for 1867, contains two interesting state-

[5] Letter in the author's possession.

The founders of the National Academy of Sciences. Left to right: W. B. Patterson, A. D. Bache, Joseph Henry, Louis Agassiz, Abraham Lincoln, Senator Henry Wilson, Admiral Charles H. Davis, and B. A. Gould. (From a painting by Albert Herter.)

Henry presiding over the National Academy of Sciences in Mineral Hall, Smithsonian Institution. 1, Joseph Henry; 2, Mary Henry; 3, W. J. Rhees; 4, F. W. Clarke; 5, J. S. Newberry; 6, J. C. Dalton; 7, J. E. Hilgard; 8, J. J. Woodward; 9, Peter Parker; 10, Alfred M. Mayer; 11, William Ferrel; 12, Benjamin Silliman; 13, C. E. Dutton; 14, Emil Bessels; 15, Arnold Guyot; 16, J. H. C. Coffin; 17, B. A. Gould; 18, Elias Loomis; 19, C. A. Schott; 20, George Engelmann; 21, Benjamin Peirce; 22, Simon Newcomb; 23, Lewis H. Morgan; 24, A. A. Michelson; 25, John S. Billings; 26, Weir Mitchell; 27, Frederic M. Endlich.

ments. The first was a disclaimer that he was one of the Academy's founders. "I feel myself more at liberty to urge the claims of the Academy, inasmuch as its members, including myself, took no steps toward its establishment. Indeed, I must confess that I had no idea that the national legislature, amid the absorbing and responsible duties connected with an intestine war, which threatened the very existence of the Union, would pause in its deliberations to consider such a proposition." [6]

This dismisses the suggestion that Henry was one of those who took an active part in the establishment of the Academy, but it does not say that he was lukewarm in his sentiment toward it. It is more probable that after Bache had broached the matter in 1851, he and Henry discussed the project in some detail, even bringing Benjamin Pierce and B. A. Gould into their discussions. It is not at all unlikely they had evolved a well-conceived plan that awaited nothing but Agassiz's enthusiasm and the ear of a sympathetic senator to set it moving toward realization.

The other statement of interest in this annual report was the definition of his desire to see the scientist receive adequate public recognition for his labor.

Indifferent he may have been to financial reward, but Henry was never indifferent so far as the reception of his work was concerned. He had a sense of his own importance that was tempered by his humility in advancing claims. We have seen how the smart of neglect had prompted him to appeal for notice to Oersted while at Princeton. But it was not merely a personal matter. He desired that scientists in general should meet with a fuller degree of recognition. From this desire sprang his suggestion to make membership of certain societies a recognized mark of distinction. Membership in the National Academy of Sciences, at first limited to fifty in number, seemed an appropriate criterion of outstanding merit.

Henry chose the first opportunity to advance his proposal.

It was implied in the organization of such a body that it should be exclusively composed of men distinguished for original research, and that to be chosen one of its members would be considered a high honor, and consequently a stimulus to scientific

[6] *Nat. Acad. Reports,* 1867. Sen. Misc. Doc. no. 106, 40th Congress, 2nd session, 1868, p. 5.

labor, and that no one should be elected into it who had not earned the distinction by actual discoveries enlarging the field of human knowledge.

... I am happy to say that in filling the large number of vacancies which have been occasioned by death and resignation since the original organization, the principle before mentioned has been strictly observed, and no one has been admitted except after a full discussion of his claims and a satisfactory answer to the question "what has he done to advance science in the line of research which he has especially presented?"

The organization of this academy may be hailed as marking an epoch in the history of philosophical opinions in our country. It is the first recognition by our government of the importance of abstract science as an essential element of mental and material progress.[7]

Bache had shared Henry's opinion that one of the principal objects to be sought in the foundation of the Academy was to afford recognition to men of achievement in science. But Henry's suggestion in the report just quoted might be interpreted to mean that election to membership would be an official badge of distinction comparable to the honors bestowed by monarchs. This proposal to create an aristocracy of intellect would be opposed as being inconsistent with democratic principles. Henry did not evade such an argument; he met it head on.

> It is not enough for our government to offer encouragement to the direct promotion of the useful arts through the more or less fortunate efforts of inventors; it is absolutely necessary, if we would advance or even preserve our reputation for true intelligence, that encouragement and facilities should be afforded for devotion to original research in the various branches of human knowledge. In the other countries scientific discovery is stimulated by pensions, by titles of honor, or by various social and official distinctions. The French Academicians receive an annual salary and are decorated with the insignia of the Legion of Honor. Similar marks of distinction are conferred on the members of the academy of Berlin and St. Petersburg.
> These modes of stimulation or encouragement may be con-

[7] *Ibid.*, p. i.

sidered inconsistent with our social ideas and perhaps with our forms of government. There are honors, nevertheless, which in an intelligent democracy have been and may be justly awarded to those who enlarge the field of human thought and human power. Heretofore, but two principal means of distinction have been recognized in this country, viz: the acquisition of wealth and the possession of political power. The war seems to have offered a third, in bestowing position and renown for successful military achievement. The establishment of this Academy may be perhaps regarded as having opened a fourth avenue for the aspirations of a laudable ambition, which interferes neither with our national prejudices nor our political principles, and which only requires the fostering care of the government to become of essential benefit and importance, not only to this, but to all the civilized countries of the world.[8]

Whatever merit these views may have borne, the government failed to respond and showed no disposition to furnish "the fostering care," and no tangible distinction has ever been associated to membership in the Academy such as that attached to judicial station, military rank, or high political position. But Henry's efforts to obtain recognition for a body of men not given to self glorification and, therefore, prone to be overlooked in a vociferous society, only served to endear him further to his fellow-workers.

Henry's claims that the scientist should receive recognition by the national authorities could not be attributed to any vain longings on his own part. He was content with the title "Professor" which he had earned by his own efforts. Recalling at a later date that he was the holder of nine offices, he was careful to remark they had "all been pressed upon me without solicitation on my part." In truth, the only office or position he ever sought was that of village schoolmaster while he was still studying at Albany.

His advocacy of the scientist's right to recognition could not have arisen from vanity, he was modest to excess. Nor did it spring from unsatisfied ambition, for Henry was devoid of ambition. If one must find a motive for his course, it will surely be found in his belief in the old commonplace that glory should follow and not be followed.

Union College had been first to single out Henry for honors when

[8] *Ibid.*, p. 3.

it awarded him the degree of LL.D. in 1829. South Carolina University followed with a similar award in 1838, and Harvard conferred this degree upon him in 1851. If in these modern days of abundant academic honors it might appear that the universities had withheld their appreciation, it should be remarked that in Henry's time honorary degrees were not distributed with generosity. Faraday was exceptionally well treated, but the great Humphrey Davy received but one honorary degree, that of Dublin.

Henry's labors on behalf of his fellow-scientists were much appreciated. The younger Silliman wrote that the long period of absence from the laboratory, "over thirty years, devoted so largely to work almost purely administrative, was a severe tax upon a man of Professor Henry's great productive power and ability in original research can hardly be questioned. It rarely falls to the lot of any man of science to do so much for the best interests of the entire body of scientific workers, or to succeed so well in securing respect and confidence on the part of the public for scientific results and methods." [9]

Henry was prominent in yet another society but his attendance at its meetings seem rather to have been a relaxation, so much joy did he derive from them. A small group of kindred spirits like Bache, Schaeffer, Saxton, General Meigs, and W. B. Taylor, all engaged in work for the government, formed the Saturday Club. This club met weekly at the residences of its members in turn, although rather more frequently at Henry's apartment, and the host often invited guests not affiliated with the club.

The city of Washington attracted an unusual number of men in scientific and technical occupations. In addition to the normal professional class in the national capital, there were the staffs of the numerous government bureaus, such as the technical branches of the Army and Navy, Coast Survey, Office of Weights and Measures, the National Observatory, the Patent Office, the Hydrographic Office, and so forth.

More and more of these men, having been guests at meetings of the Saturday Club, expressed a wish to become regular members. Thus it was decided to adopt a more comprehensive title, a constitution, and by-laws. Under the title of the Washington Philosophical Society the new organization held its first meeting on March 12, 1871, at the Smithsonian Institution. Henry was in the chair.

[9] *Amer. Jour. Sci.* 1878. Vol. CXV, p. 468.

ORGANIZER OF SCIENCE

He took a keen interest in this organization, of which he was the first president. He was foremost with remarks upon the papers presented. Although well-advanced in years, instead of suffering from the defects of age, his mind was unclouded and his qualities unimpaired. His mind was a storehouse of scientific knowledge, and he had the happy faculty of illuminating his knowledge by flashes of private wisdom.

After his death this society decided to preserve his memory by establishing the Joseph Henry Memorial Lecture. The first of these lectures was given by Joseph S. Ames, president of Johns Hopkins University, in the year 1931, when the entire world was celebrating the centenary of the discovery of electromagnetism.

JOSEPH HENRY, THE MAN

Joseph Henry achieved and maintained distinction as much by his character as by his attainment. We know Henry chiefly through his published works, and these relate almost exclusively to his scientific activities. Because we have no personal journals and have been largely deprived of his personal correspondence, the man is revealed to us only in a patchwork of fragments. We are thus denied most of those intangible human values by which the plenitude of a man's character can be judged. To a great extent we must judge him as he is, to quote Paul Valery, "regulated by his own powers of thought." He outlived most men of his own generation, and most estimates of the man that have come down to us are made by men of a younger generation.

Any description of his personality must of necessity be based upon the comments of those who knew him only in later years. Because he refrained from self-expression and restricted his writings only to his work, it is easy to fall into the erroneous view that he was interested only in science and neglected human associations. Such a judgment cannot be true. We have many assurances that he possessed a warm nature which endeared him to his students, fellow scientists, and many others outside the field of science.

Henry's life is without the great emotional or moral conflicts that make interesting reading in novels. His inaction in the matter of square or round-toed boots, or in the matter of publication of his research, betrays a kind of characteristic indecision, but he was never in doubt about a basic philosophy of life. It was a placid life, full of the delights of family and friends, filled with stimulating and satisfying work.

JOSEPH HENRY, THE MAN

We have followed him to old age, when, to most men, the dreary drift of the years becomes apparent. It is the time for repose, for retirement from the strain and fret of routine existence. Strong souls like Henry do not take rest easily. Indifference weighs them down; they demand a mission, a motive for action and faith. It was Henry's good fortune to have this mission.

He was at the height of his reputation, dean of American science, and head of a great international institution. A distinguished contemporary said of him:

> Professor Henry was a man of whom it may be said he held a place which has never been filled. I do not mean his official place, but his position as the recognized leader and exponent of scientific interests at the national capital. A world-wide reputation as a scientific investigator, exalted character and inspiring presence, broad views of men and things, the love and esteem of all, combined to make him the man to whom all who knew him looked for counsel and guidance in matters affecting the interests of science. Whether anyone could since have assumed his position I will not venture to say; but the fact seems to be that no one has been at the same time able and willing to assume it.[1]

Of his appearance we have only the casual words of acquaintances and the more revealing portraits. From these it is possible to reconstruct a picture.

We have the sharp impression of a well proportioned form, over six feet in height, and a straight line from shoulder to heel. He had that upright solid kind of dignity that reminds us of one of his lighthouses. His carriage, and something in his manner, something that attracted and yet held at arm's length, gave a hint of mystery to the man. We perceive at once that he must have been handsome in his younger days, but in most portraits he is old, past seventy, but with the forceful head of a man of action gifted with the faculty of deep thinking. For there is no mistaking the depth of mind behind those eyes.

The wrinkles of his face are not those of the social smile, they are the wrinkles of sunshine, for the gentle mouth is curved in a smile of ambient benevolence which endeared him to his associates and caused them to forgive his little foibles.

[1] Newcomb, S. *Reminiscences*, p. 234.

If Henry's face says anything to the beholder, it warns the world it has to deal with a man of perfect self-control. The face we see in the portraits is not that of a simple scholar of the study, but rather of a man with a brain. It was an acutely organized brain, as well-balanced as it was keen.

He did not speak idly. If he had anything to say he said it, but if he had nothing to say it was no hardship to preserve silence. He was probably a self conscious man; by holding his tongue he contrived to conceal this weakness. The fault, if it was a fault, cannot be attributed to egotism, for egotism is really a failure of sympathy or a failure of proportion, and of these faults Henry cannot be accused. He was a man of ready sympathy, and his self-consciousness probably arose from shyness.

His protective reticence may merely have been that of the ordinary sensitive and modest man. It is prone to be misunderstood by those who do not come into intimate contact with the men who possess these qualities. There were some who thought his reserve in conversation arose from frigidity, and who regarded him as unimpressionable, but these were few, and either were casual acquaintances or required something from him he would not give. Of those people we shall have more to say later.

When the time came to speak out, he did so fearlessly. His bold confrontation of Senator Douglas in a public meeting proved that he did not lack moral courage. In private he could be so frankly outspoken as to utterly disconcert his listener. There is in existence in his handwriting a memorandum of a conversation he had which reads:

> Mr. R. [probably Renwick, the architect] stated that he had not been informed of the Resolution of the Board relative to the payment of the contractor. He then went on to state what conversation he had held with Cameron and to prove to me that he could not have known of the Resolution. I said I did not believe it. Then, said he, you doubt my word and either you are a liar or I am. Then, said I, if you put it on that ground, you are the liar.

He was a proud and sensitive man and, as Voltaire has told us, pride does not produce laughter. Henry does not appear to have had any sense of humor, although some of his friends claim that he only refrained from enjoying the humor of others when it bore some cruel barb.

Henry was occasionally misunderstood because he avoided the crowd. He never sought popularity. It was fortunate that he did not do so, for he had some qualities that make for greatness but few of the lesser graces which attract an easy popularity. At first, concentration on his laboratory work and, later, absorption in his administrative duties did not leave him the leisure to cultivate relations with the multitude. Except at scientific gatherings he rarely appeared as a public speaker. Constantly in demand to make speeches and addresses, he never seems to have contemplated following the example of the elder Silliman, who became a famous popular expositor of the sciences.

Henry is not remembered for witty or profound sayings in either his speech or his writings. He was not a phrasemaker. His writing was like his speech, clear, restrained, and forceful rather than sensitive. The nature of his writings did not preclude the employment of a literary style, but there is no urbanity, no verbal agility in what he wrote. Working as he did in the initial stages of some branches of science which had not yet developed a vocabulary, he had to describe his observations in a short and simple way but, if his writing lacks the grace of water playing from a fountain, it at least flows clean and granular like fine dry sand.

He did achieve in his lectures and papers an incomparable lucidity which shows the shape of the thought within, if it never ascends to the height of thought taking shape in a perfectly pure medium of language. Coherent and connected, it is simple workaday prose that does not demand a rereading to grasp the meaning.

To those with whom he worked daily he was attached by bonds far stronger than those who are thrown into business association usually develop. He showered unobtrusive kindness upon those around him. In assuming his position as Henry's successor as secretary of the Smithsonian Institution, Professor Baird spoke of his deep attachment to his former chief, and his daughter bears witness to the genuineness of the feeling. She wrote: "The relations between Professor Henry and my father were of the most friendly character, Professor Henry's kindness growing with every year of their official association and my father's affection for his chief increasing steadily until finally the feeling became almost fatherly and filial between the two. . . . His admiration for Professor Henry's genius for the noble work he did was deep and enthusiastic." [2]

If Henry had shortcomings, as every man must, they are dwarfed beside his virtues.

Genius does not appear to the best advantage when seen in shirt sleeves and carpet slippers. Henry has been spared from this exposure because we have few intimate pictures of him, and his character rarely flowed into incident.

His home in the east wing of the Smithsonian building was an acknowledged center of intellectual life in Washington, where cold intellect was tempered with warm hospitality. There were few distinguished visitors to the capital who did not visit him, yet there are no records of the many meetings of minds which must have taken place within those rooms.

His home was a refuge. If we have no pictures of the interior it is because life within was of that domestic happiness which flourishes only in privacy. A devoted and understanding wife sheltered him from the thousand pin-pricks that may distract a man in public office. Although there must have been many occasions when the necessity of living within his narrow means could have been no simple feat, he was spared most of the fret and irritation which accompanies a straitened domestic economy.

This he appreciated. In return he allowed his sunny nature to unfold at the fireside. The little family would talk over his work, and it was rare when the felicity of their circle failed to disperse the shadows that sometimes found entrance to his thoughts. Quick to offer sympathy, he was no less quick to respond to its influence.

The observation which is applied intensely to the analytic study of nature is often dulled to the perception of beauty and poetry in the wide sweep of nature. This was not true of Joseph Henry. It was no maudlin sentiment that brought tears to his eyes when he came unexpectedly upon a scene of great natural beauty. He was a humble and sincere admirer of Creation who sought its pattern in the flash of lightning and the electric spark, no less than in the monumental peace of the mountains or in the abiding silence of the forest.

The joy of these sights are denied for much of the year to men who

[2] Dall. *Baird,* p. 394.

work within four walls. During their hours of relaxation, the wise seek to catch glimpses of beauty through the eyes of the poets. Henry was one of these. He was, moreover, a frequent visitor to the gallery which Mr. Corcoran had filled with pictures in his mansion on H Street, and he was such an appreciative visitor that he was appointed one of the trustees when the gallery was handed over to the nation. He spent several joyful hours in the studios of painters, where he was just as much at ease before a canvas as he was in his own apparatus room.

Best of all was the solace he derived from the measured music of poetry. The rhythmic message rang true to his sensitive mind. When there was nothing new to claim his attention he could take refuge in the reassuring shelves of the past. His strong nature worship and the kinship of their philosophy gave him a fondness for the poets of the Romantic Revival, like Southey, Coleridge, and Wordsworth. While in London, he had heard the latter lecture on poetry, and had enjoyed the experience, although he could not recall a great deal of what the poet had said. He devoured Shelley, who felt the cosmic ecstasy more than any other English poet.

But this was reading for the contemplative mood. There were other authors who looked less deeply into the sublimities but who shared more generously of homespun human sympathies, and these, too, received a fair share of his appreciation.

Henry thoroughly enjoyed the evenings spent with his family, when the fire leaped in the chimney and the lamps were lit, and the wind moaned outside, and the contented mind possessed its dreams. This was the time when the man whom the world knew laid aside his cares and became husband and father. After one of the daughters had entertained them at the piano, another would begin to read aloud. His daughter Mary briefly admits us to the intimacy of these domestic scenes.

> Had father with us all the evening. I modelled his profile in clay while he read Thomson's *Seasons* to us. In the earlier part of the evening he seemed restless and depressed but the influence of the poet drove away the cloud, and then an expression of almost childlike sweetness rested on his lips, singularly in contrast yet beautifully in harmony with the brow above.

And on another occasion:

> We were all up until a late hour, reading poetry with father and mother, father being the reader. He attempted *Cowper's Grave* by Mrs. Browning but was too tender hearted to finish the reading of it. We then laughed over the *Address to the Mummy,* soared to heaven with Shelley's *Skylark,* roamed the forest with Bryant, culled flowers from the poetical fields, and ended with *Tam o'Shanter.* I took for my task a part of the latter from memory, while father corrected, as if he were "playing schoolmaster." [3]

Henry often astonished his acquaintances with the depth and range of his reading, but he had an especially intimate knowledge of poetry, and he would frequently cap someone's words by delving into his memory to produce an apt passage. He seems to have been aware of the close relationships between the truths of science and of poetry, with the poet having the advantage because he could mingle beauty with truth. Perhaps this was because Henry felt truth rather than reasoned his way toward it, realizing that some truths are not discernible through controlled experiments, that these truths are only perceptible through the eyes of the mind.

He never entirely outgrew his early taste for romance. He chose a novel to furnish relaxation after a period of continuous and hard thinking. He kept in touch with the past by reading, and he possessed the faculty of using history as the key to understanding the problems of the present. He was not so interested in the historical fact itself as in its relationship to the several facts that made the basis for generalizations. The civilization of ancient Greece was a source of never failing interest to his speculative mind. He was unusually well read in the literature, art, philosophy, and polity of the early city states, and found joy in the classic writers of Greece and Rome.

Few men have had a wider circle of friends. He had friends among statesmen, like Lincoln and Davis; in law and politics, like Calhoun, Taney, and Chase; in the services, like Sherman and Meade; in the civil service, like Chittenden and Hilgard; but especially in science and scholarship he had friends innumerable. He looked upon all men with friendliness and understanding, and some response was

---

[3] Welling, J. C. Life and character of Joseph Henry. *Memorial,* p. 193n.

won from the most angular. He had a Tennysonian faith in human nature which caused him to love humanity, not as an abstraction, but as a community of lovable and fallible human beings. It is not difficult to understand why Henry should have compelled a response from other men. Dr. Samuel Dod, a Princeton colleague, summarized the source of attraction: "To those who knew Professor Henry personally, there was the charm of a singularly gentle and unaffected sincerity of heart and manner, that made him approachable to all. His attachments were warm and lasting." [4]

Those who enjoyed an intimacy with him acclaimed his wisdom and ripeness of spirit. Asa Gray testified: "In the evening we fell to discoursing on philosophical topics, and Henry threw out great and noble thoughts, and as we both fell to conversing with much animation my headache disappeared entirely. There is no man from whom I learn so much as Henry. He calls out your own powers, too, surprisingly." [5]

It is not easy to draw on a little canvas the picture of a man whose nature is large and central, without cranks or oddities. The simplicity and wholesomeness of such a creature defies an easy summary, for he is too spacious in his effects. So long as his friends lived he was held in affectionate remembrance, for it was not simple to overrate his goodness, and none was found ready and willing to raise a voice in criticism or disparagement.

He was rare because he was so superbly normal, so wholly in tune with the world and with other men. Whatever cranks or mannerisms he possessed, and there must have been idiocyncrasies in his character, must have been of the minor sort, for they did not remain in the memory of those who wrote recollections of him.

A life of domestic happiness, innumerable friends, a world enlivened and comforted by his presence—what more can mortal man wish for?

Henry's religious beliefs occupied a prominent part not only in his thoughts, but in his way of living. He displayed the power and sincerity of his convictions by making his life conform to the principles he professed. This is evident in the simplicity of his life, the

---

[4] Memorial discourse. *Memorial,* p. 142.
[5] *Letters.* Vol. I, p. 349.

rectitude of his actions, and the active benevolence which characterized his relations with lesser men.

To attribute, as Crowther does, Henry's austerity to the influence of his mother's puritanical beliefs is a misplacement of emphasis, it overlooks the fact that these traits are to be observed in the character of other scientists who have not incurred the charge of having been brought up in a religious strait-jacket. Moreover, it is to ignore the influence of his time and occupation. Most men of his time with a practical turn of mind sought employment in those fields where their wares found a market. The men whom Henry followed and those who were his co-workers took no delight in the flesh-pots. They called themselves philosophers, and their declared purpose was to learn the truths of nature.

It would be wrong to suppose these men were incapable of enjoying life because their chief joy was in their discoveries. The overtones of their delight in accomplishment were restrained, but they can be discerned beneath the measured gravity of their writings. Fortunately for science, this spirit has never died out. It happens to have been strong in Henry's character and it was not submerged under his efforts to gain recognition for himself and his fellow-workers.

Henry was not a religious man in the common sense. Although he was one in whom the soul drew near the surface of life, there were no groaning hours of communion with the spirit. Ardent followers of Calvin and Knox found adherence to a fixed creed their highest good; values were not to be thought about and analyzed, merely accepted. If Henry had to fight any stormy battle with his Calvinistic background, he came out the victor, as his later life shows.

A Puritan he may have been. But there is dignity and discipline and majesty about puritanism, and the best of puritans have found refuge in abasement, if only before God. He was, like Faraday, a scientist and Christian. These two men, alike in so many matters, were akin in their ability to reconcile their science with their religion. Although Henry lived as one who has a continuous vision of the unseen, he would surely have subscribed to the doctrine which ascribes a dual function to science—to acquire for man an accurate picture of the world of phenomena and to provide him with a means of controlling his environment. To this statement could be appended the caution that to proceed further is to exceed the bounds of science. "Man, with his finite faculties, cannot hope in this life to arrive at a knowledge of absolute truth; and were the true theory of the universe, or in other words, the precise mode in which the Divine Wisdom operates in producing the phenomena of the material world, revealed to him, his mind would be unfitted for its reception; it would be too simple in its expression, and too general in its application, to be understood and applied by intellects like ours." [6]

These words reflect the attitude of another distinguished scientist who influenced Henry's thought, Hans Christian Oersted. In the year 1849, Oersted had published his *Aanden i Naturen,* a philosophic work of no little interest at the time. Translated into German it attracted many readers. Henry knew the work, and it touched him so deeply that he contemplated translating it into English, but he never found time to do it.

The book is a series of philosophic treatises, popular in treatment, to serve as an introduction to nature by revealing the internal spirit which determines all its phenomena, and the relation in which the spirit exists toward the material and intellectual worlds. There seems little room for doubt that the minds of Oersted and Henry were in harmony, especially upon the view of the spirit manifested in relation to matter. They agreed that there was a constant and immutable essence in the continual changes in nature, as though a single thought or design existed. The unity of this thought pertains to nature, for the natural laws, which are constant and unvariable, are laws of reason. But not the reason which is within us. Rather is it a reason which prevails throughout the universe.

When Darwin's *Origin of Species* was published in America in 1860 it provoked the same storm of criticism and opposition it had aroused in England in the preceding year. The book divided science against itself, but that was a division which might be expected to close gradually in the course of calm discussion. More strident and vociferous was popular opinion supported by religious views.

The mutability of species was not a new doctrine. Buffon and Lamarck abroad had more than hinted at it. In this country, Raffinesque, an able botanist, and Samuel Haldeman, naturalist and philologist, had advanced the thesis. But Darwin's biological proof coming to support that of Lyell the geologist, disregarding Biblical chronology, asserting the antiquity of man upon the earth, and rejecting the

---

[6] On the Theory of the So-called Imponderables. A.A.A.S. *Proceedings.* Vol. VI, p. 87; *S.W.* Vol. I, p. 300.

belief that each species of living thing was the result of an original divine act, was too much for orthodox Biblical students.

Battle was immediately joined. While the conflict raged, many people who were genuinely perplexed and unable to furnish their own answers to the questions raised, waited expectantly for an expression of opinion by Henry. He was known to be a devout Presbyterian, regular in attendance at church worship, so that he could be relied upon to approach the subject with a reverent attitude. He would not lend a hand in destroying what many generations had built merely because some of the plasterwork was shaky.

Had he been held in the strait-jacket of austere Calvinism now was the time to uphold the orthodox beliefs with all his eloquence.

He did nothing of the kind. He maintained a sphinxlike silence.

Henry's friend Asa Gray, the foremost field botanist in America, had been in Darwin's confidence before the book had been published. He was the champion of the theory of evolution in this country and became embroiled with dissenting scientists and protesting churchmen. He found an unexpected supporter in Henry, who wrote to him: "I have given the subject of evolution much thought, and have come to the conclusion that it is the best working hypothesis which you naturalists have ever had." [7]

As usual, Henry had taken his time before arriving at any conclusion. He had been turning the matter over in his mind for four years before he made this concession.

When he communicated this view to Gray it became known to his other friends at Harvard. America's most brilliant biologist, Louis Agassiz, learned of Henry's attitude with dismay. How or why Agassiz, with all his brilliance, could reject Darwin's theory of evolution surpasses understanding. Few men had done more to provide the evidence upon which the theory was grounded. However, he held tenaciously to the idea that there were periods of creation at intervals during the various successive geological ages, that all the species of animals and plants created at each period lasted for a given time in order to be replaced successively by others.

There was a deep attachment between Henry and Agassiz. Asa Gray once found them in his home awaiting his return. "Agassiz and Henry enjoy and admire each other so richly," he wrote, "and talk science so glowingly and admiringly, that I think I should not have

[7] Goode. *The Secretaries*, p. 150.

been at all surprised to see them exchange kisses before they were done." [8]

Learning that Henry was sympathetic toward Darwin's views, Agassiz decided he must take action. The unique position Henry occupied in contemporary thought would make him the most formidable accession of strength to the evolutionists. Agassiz believed it was too hazardous to permit Henry to issue any statement supporting Darwin. He went to Washington in order to undertake his friend's conversion. This was an extraordinary undertaking, for he must have known with what painstaking care Henry reached his decisions, and once reached, how tenaciously he held to them. Agassiz failed in his mission to make a convert but he pleaded with the sage of the Smithsonian to abstain from making any public statement, either for or against Darwin's views. Agassiz loved the public platform, he revelled in controversy. Henry shunned all forms of publicity and had a horror of public debate. It could not have been difficult to persuade him to be silent.

The remonstrances of Agassiz were supported by the members of the New York Avenue Presbyterian Church with which Henry had been associated for the twenty years he had been in Washington, and of which he was a ruling elder. They implored him not to undermine the faith of his coreligionists who did not have his superior knowledge upon which to support themselves.

Henry never did take the public into his confidence on this controversial subject. His silence may be open to criticism. If everyone who had formed opinions abstained from expressing them out of deference to the people who were not mentally prepared for new ideas, there would be an end of all progress.

If Henry made no announcement of his views we may be sure that it was not because he was swayed by remonstrance. The simple truth was, he would not under any circumstance give expression to opinions while he lacked assurance that he possessed all the facts. He was a physicist, not a naturalist, biologist, or theologian. The length he had gone in declaring that Darwin had furnished the naturalist with his best working hypothesis was as far as his knowledge would take him. Even this attitude was something of a sacrifice, since it aligned him against such treasured friends as Torrey, Guyot, and Agassiz. He would venture no farther, but neither would he retreat.

[8] *Letters.* Vol. I, p. 349.

Since he knew that much would be expected of him by both contestants and spectators in the dispute, he maintained a discreet silence. Once he decided to speak out there could be no retreat, no matter what injury he might do. While others waited for him to speak he dug deep down in his mind. He required nearly eighteen years to think out his position in the conflict between science and religion. He told his friend Welling he would discuss the relations of science and religion, and also the import of prayer in his final address to the Washington Philosophical Society but, as had happened before, he delayed too long. Before he had prepared his paper, the Great Reaper had done his work.

Even after his death Henry's friends feared that this man, poised on a height and seeking a glimpse of the promised land, might have written his views upon the subject, and that posthumous publication might harm the church by undermining the faith of its adherents. They knew him to be fearless. No one doubted that if he took a stand it would be uncompromising.

Simon Newcomb says of him: "My talks with Professor Henry used to cover a wide field in scientific philosophy. Adherence to the Presbyterian Church did not prevent him being as uncompromising an upholder of modern scientific views of the universe as I ever knew." [9]

So, in fearful silence, his friends waited for his pronouncement upon the subject of science and religion. After his death, it was learned that he had drawn up a form of credo, and that he had entrusted it to a friend. Henry knew at the time he wrote this letter that he did not have long to live. He died within a month after writing this simple statement of belief to his old friend Joseph Patterson, the Philadelphia banker.

> We live in a universe of change; nothing remains the same from one moment to another, and each moment of recorded time has its separate history. We are carried on by the ever-changing events in the line of our destiny, and at the end of the year we are always at a considerable distance from the point of its beginning. How short the space between the two cardinal points of an earthly career!—the point of birth and that of death; and yet what a

[9] Newcomb. *Reminiscences*, p. 408.

universe of wonders is presented to us in our rapid flight through this space! How small the wisdom obtained by a single life in its passage, and how small the known when compared with the unknown, by the accumulation of the millions of lives, through the art of printing, in hundreds of years! How many questions press themselves upon us in the contemplations whence come we, whither are we going, what is our final destiny, the object of our creation?

What mysteries of unfathomable depths environ us on every side! But after all our speculations, and an attempt to grapple with the problem of the universe, the simplest conception which explains and connects the phenomena is that of the existence of one Spiritual Being—infinite in wisdom, in power, and all divine perfections, which exists always and everywhere—which has created us with intellectual faculties sufficient, in some degree, to comprehend His operations as they are developed in Nature by what is called "Science."

This Being is unchangeable, and, therefore, His operations are always in accordance with the same laws, the conditions being the same. Events that happened a thousand years ago will happen a thousand years to come, provided the condition of existence is the same. Indeed, a universe not governed by law would be a universe without the evidence of an intellectual director.

In the scientific explanation of physical phenomena we assume the existence of a principle having properties sufficient to produce the effects which we observe; and when the principle as assumed explains, by logical deductions from it, all the phenomena, we call it a theory. Thus we have the theory of light, the theory of electricity, etc. There is no proof, however, of the truth of these theories, except the explanation of the phenomena which they are invented to account for.

The proof, however, is sufficient in any case in which every fact is fully explained, and can be predicted when the conditions are known. In accordance with this scientific view, on what evidence does the existence of a creator rest?

First. It is one of the truths best established by experience in my own mind, that I have a thinking, willing *principle* within me, capable of intellectual activity and of moral feeling.

Second. It is equally clear to me that you have a similar spiritual principle within yourself, since when I ask you an intelligent question you give me an intellectual answer.

Third. When I examine the operations of Nature I find everywhere through them evidences of intellectual arrangements, of contrivances to reach definite ends, precisely as I find in the operations of man; and hence I infer that those two classes of operations are results of similar intelligence.

Again, in my own mind, I find ideas of right and wrong, of good and evil. These ideas, then, exist in the universe, and, therefore, form a basis for our ideas of a moral universe. Furthermore, the conceptions of good which are found among our ideas which are associated with evil, can be attributed only to a Being of infinite perfections, like that which we denominate "God." On the other hand we are conscious of having such evil thoughts and tendencies that we cannot associate ourselves with a Divine Being, who is the Director and Governor of all, or even call upon Him for mercy, without the intercession of one who may affiliate himself with us.

I find, my dear Mr. Patterson, that I have drifted into a line of theological speculation; and without stopping to inquire whether what I have written may be logical or orthodox, I have inflicted it upon you.

Please excuse the intrusion, and believe me as ever,

Truly yours,

JOSEPH HENRY

Thus it will be seen that Henry still clung essentially to the theology of his Puritan ancestors, dulcified somewhat by time and experience. As Crowther has wisely pointed out, a theology which ascribes equal importance to determinism and will is one of the least unfavorable to the progress of science, and that the rise of science has been intimately connected with this same theological idea.[10]

Henry's abstention from seeking the financial rewards which would have been derived from patents (an example set for him by Benjamin Franklin) was not the only evidence of deliberately turning his back on material gain. His work for the Light House Board and for other government departments was performed without any remuneration.

[10] Crowther, J. G. *American Men of Science*, p. 220.

JOSEPH HENRY, THE MAN

He was never in the pay of the government. He even declined to accept for his own use the fees he received for lecturing. This money was paid into the Smithsonian account.

When he was voted an increase in salary as secretary of the Institution, he declined to accept it on the grounds that what he received was proportionate to the sums paid to other members of the staff.[11] All must be placed upon a higher scale of remuneration; he could not consent to be singled out for favorable treatment.

Speaking about financial matters toward the end of his life, he said: "I resolved at an early age to preserve my independence by never expending more than my income, and since I have been in public life I have studiously avoided accepting propositions which have been made to me to lend my name to the advocacy of any enterprises of a speculative nature, or to accept the offered means which have been presented to me, for acquiring property, but which might compromise my independence in regard to public acts." [12]

He lived through the decade which Minnigerode disparagingly described as the "Fabulous Forties" without succumbing to its spirit. It was a time of spiritual disquiet, but no disillusion. That came later. He was equally remote from the turbulence of the decade which followed the termination of the Civil War. A good deal of strength of character was called for in resisting the golden opportunities offered by association with the manufacturing, the extractive industries, and transportation. Nineteenth-century capitalism with its emphasis upon competitive individualism was tightening its grip on the nation. The close pursuit of the narrow individual interest, the tendency to judge a man's worth by the money he made, was totally alien to Henry's ideas of society.

Some may look upon the work of Joseph Henry and scientists like him and find, in all sincerity, many short-comings because it was not "practical," so it becomes imperative to point out that to Henry science was not a practical undertaking. There are many views of science, but any which fails to embrace its meaning as a progressive development of conceptual schemes and methods and techniques by which to test and advance them cannot express the spirit of the scientist like Henry. He did not regard the technology, on which such

---

[11] In 1865, his salary was raised to $4,500, and the suite of seven rooms on the second floor of the East Wing was allotted to him for life.
[12] Letter to Robert Patterson. *Philadelphia Public Ledger.* May 14, 1878.

high value is set, and which in part, motivates science, gives it tools, and, in another part, draws upon it for its own advances, as being science.

By embracing this philosophy Henry deliberately refused to undertake the work which would bring him wealth, but, accepting it as the better part, he pleaded for recognition of the scientist so that he might receive a fair share of the material rewards of life. What he saw in advancing this plea was the scientist who had to combat ignorance and stupidity, who had to endure privation while engaged in performing a thankless toil. What he failed to see in the situation was that much of the frustration was attributable to the scientist's failure to realize his social obligations. Not all the fault lay with the public, although it must be conceded that men everywhere were ethically unprepared for the great bounty which science and technology were producing.

Henry's renunciation of wealth isolated him from the capitalistic stream which had become the main current of social life. Had he, like his British contemporaries Davy, Faraday, and Kelvin, who supplemented their incomes from outside sources, accepted the offer of fees from manufacturers and company promoters for giving his approval to their schemes, he could have ridden the tide of social success and have become something of a popular national figure, as did Edison and Morse.

Although reluctant to grasp at the financial rewards, Henry was not completely free from the springs of human ambition. His letter to Oersted lamenting the failure of European scientists to accord him the credit he felt was his due, and his work for the learned societies and for the prestige of scientists, points to a feeling that the esteem of one's intellectual peers was a distinction to which a man might justly aspire. He did not despise applause when it came from the right source.

Nevertheless, the temptation at times must have been strong. He saw men with whom he was constantly in contact, the framers of public policies, judges and lawyers, occupying conspicuous positions in public life and social esteem, yield to the temptation of making wealth in ways that were not altogether innocent. More than once he had the opportunity to endorse schemes by which others made large sums of money, but in each case he not only refused to join in deceit, but he denounced the frauds.

Novel industries were springing up with the progress of invention and these had to compete with the projects of pseudo-scientific humbugs who sought to profit from the spirit of wild speculation which had gripped the nation. Both legitimate and illegitimate promoters would have given much to have the support conveyed by the authority of Henry's name.

There is in the Smithsonian a letter from a firm which states that they were negotiating for the contract "for coating the telegraph lines" between Jersey City and Philadelphia. The writer offers Henry $100 if he will give an endorsement of the firm's product. In the margin, Henry has written, "Bribery—angry reply." The reply was sufficiently angry to produce a contrite apology. Henry summarized his disapprobation of the materialism of the times while speaking on the tendencies of education:

> A result of the widespread diffusion of elementary knowledge without a proper cultivation of the higher intellectual faculties, and an inculcation of generous and unselfish principles, is the inordinate desire for wealth. To acquire power and notoriety in this way requires the least possible amount of talents and intelligence, and yet success in this line is applauded, even if obtained by a rigid application of the dishonest maxim that "all is fair in trade. . . ."
>
> No one who has been called upon to dispense public money can have failed to be astonished at the loose morality on the part of those who present claims for liquidation. The old proverb is here generally applied, namely, "the public is a goose, and he is a fool who does not pluck a feather." [13]

This steadfast aloofness from the ordinary affairs of life did not prevent the more discerning men, those in closest touch with realities, from esteeming Henry's qualities. J. C. Welling, the president of Columbia University, wrote:

> The fascination of Professor Henry's manner was felt by all who came within the range of his influence—by men with whom he daily consorted in business, in college halls, and in the scientific academy; by brilliant women of society, who, in his gracious

[13] Thoughts on Education. *Jour. of Education.* 1855. Vol. I, p. 17.

presence owned the spell of a masculine mind which none the less was feminine in the delicacy of its perceptions and the purity of its sensibilities; by children who saw in the simplicity of his unspoiled nature a geniality and a kindliness which were akin to their own. . . . It was the breadth and catholicity of Henry's intelligence which enabled him to find something unique and characteristic in persons who were flat, stale, and unprofitable to the average mind.[14]

Henry's services were given generously to the nation's leaders; he served in a broad fundamental sense industry, agriculture, and marine transportation, but he declined consciously to serve any one class. He served all, however different their ideals. Had he chosen to favor any one class he would not be the half-forgotten figure he is. He might have become something akin to a patron saint.

Presidents he served indiscriminately, not because he admired them, but from a sense of duty, and not because he conceived of government as representing a dominant group. This has interfered with an appreciation of his greatness. The voluntary isolation from the main current of his times prevented the nation from realizing the stature of the man. They could not understand him, nor he them. He began and continued as a scholar-scientist, and he ever retained his loyalty to that group of men.

A professional tradition has since grown up forbidding the scientist from patenting his findings for private gain, and scientists have been slow to take financial advantage of their new knowledge. It is a tradition founded upon the examples of such men as Franklin, Henry, Faraday, Agassiz, and Pasteur. But one may entertain doubts whether there is any special virtue in too rigid an observance of these rules. A man may follow science in the pursuit of truth, or in the hope of finding wealth alongside of truth without in any way lowering his value in society.

It was inevitable that Henry's reputation for integrity should appeal to those who had reason to seek its aid in pursuit of shady designs. His first contact with the numerous swindles that became prominent during this period of commercial turmoil began in 1854 with a project

[14] Life and character of Joseph Henry. *Bull. of Phil. Soc. of Washington.* Vol. II, p. 203.

for separating alcohol from water by the stress of their specific gravities when suspended in a long column. A patent was obtained for this new method of rectifying spirits, much favorable publicity was secured, and all that remained for the production of "mature" whiskey and brandy within a few hours was a public convinced that here was a sure way to make a fortune.

Henry strangled this deception by a single test. One of the Smithsonian towers was converted into a laboratory. "A series of stout iron tubes of about an inch and a half internal diameter formed the column, the total length of which was one hundred and six feet. Four stop cocks were provided, one at the bottom, one about four feet from the top, and the other two to the intermediate space equally divided or nearly so." [15]

Accurate hydrometer readings were made over a period of six months, a portion of the liquor in the column being subjected to comparative tests with a quantity held in reserve. The results were disappointing to the prospective distillers.

"There is not the slightest indication of any difference of density between the original liquor and that from the top or bottom of the column, after the lapse of hours, days, weeks, or months. The fluid at the bottom of the tube it must be remembered was for five months exposed to the pressure of a column of fluid at least a hundred feet high."

Robert Keeley was one of the more picturesque figures who pestered Henry for a certificate of approval for his wild schemes. Keeley was no modest improver of science. He went into the business on a grand scale by claiming to have made revolutionary discoveries in the forces of nature. He had to invent a new vocabulary to explain the startling revelations made in his laboratory.

Having made a device by which he claimed he could release the atomic energy concealed in a small quantity of water, this plausible scoundrel wheedled credulous people into investing millions of dollars in a motor that was supposed to operate by sympathetic vibrations in nature. Actually, he made a fantastic combination of parts perform a sort of motion by compressed air stored in a hidden container and fed through hollow tubes which he declared to be solid rods.[16]

Keeley was a man of modest deportment. His manner of simple

[15] A.A.A.S. *Proceedings*, 1855, p. 142.
[16] Keeley Papers in The Franklin Institute.

living and glib speech enabled him to deceive many people who, lured by the promise of making large fortunes, entrusted their savings to him. He was not content to mulct the meek, the lowly, and the ignorant. He aspired to involve such prominent men as Commodore Vanderbilt in the fraud. To gratify his ambition, he felt the need for scientific support for his schemes, and he appealed to Henry for a testimonial of approval. The grandiloquent vocabulary adopted to describe the rhythmic sympathetic vibrations of nature failed to shake Henry, who might have given serious consideration to a modicum of truth expressed in comprehensible language. He declined to enter into the deception. Instead, without mentioning Keeley by name, he dismissed the impostor's pretensions in no uncertain terms.

> We do not hesitate to say that all declarations of the discovery of a new power which is to supersede the use of coal as a motive power, have their origin in ignorance or deception, and frequently in both. A man of ingenuity in combining mechanical elements, and having some indefinite scientific knowledge, imagines it possible to obtain a certain result by a given combination of principles, and by long brooding over his subject previous to experiment, at length convinces himself of the certainty of the anticipated result. Having thus deceived himself by his sophisms, he calls upon his neighbors to accept his conclusions as verified truths; and soon acquires the notoriety of having made a discovery which is to change the civilization of the world. The shadowy reputation which he thus acquires is too gratifying to his vanity to be at once relinquished by the announcement of his self-deception; and in preference he applies his ingenuity in devising means by which to continue the deception of his friends and supporters, long after he himself has been convinced of the fallacy of his first assumptions. In this way, what was commenced in folly generally ends in fraud.[17]

Keeley was but one of those who importuned Henry for recognition. Most of them harassed him with their correspondence, and he had enough on his hands with those correspondents who were worthy of his attention. "The most troublesome correspondents," he complained, "are persons of extensive reading, and in some cases of con-

[17] *Smithsonian Annual Report for 1875*, p. 39.

siderable literary acquirements, who in earlier life were not imbued with scientific methods, but who, not without a certain degree of mental power, imagine that they have made great discoveries in the way of high generalizations. Their claims not being allowed, they rank themselves among the martyrs of science, against whom the scientific schools and the envy of the world have arrayed themselves. Indeed, to such intensity does this feeling arise in certain persons, that on their special subjects they are really monomaniacs, although in others they may be not only entirely sane, but even evince abilities of a high order." [18]

Anyone who has been associated with a scientific institution will recognize that Henry was dealing with a type of mind rather than a phenomenon peculiar to his period or his position. He had an infinite patience with the earnest seekers after truth, whether they were scholars who had made some headway in their profession, or irrepressible searchers after perpetual motion, or wayward mathematicians intent on squaring the circle, or simple citizens whose limited knowledge caused them to waste hours of hard-earned leisure upon an exploded error. He was never known to upbraid or to ridicule honest men, however foolish their pursuit appeared to be. If it were honest work or study, Henry would always find time to explain in simple language the operation of nature's laws, so that the errant might be set on the right path and their efforts not wholly wasted.

"He was accessible to a fault," said Newcomb, and "the only subject on which the writer ever had to express to him strong dissent from his views was that of the practicability of convincing 'universe-makers' of their errors. They always answered with opposing arguments, generally in a tone of arrogance or querulousness which deterred the modest Henry from replying further; but he still considered it his duty to do what he could toward imbuing the next one of the class who addressed him, with correct notions of the objects of scientific theories." [19]

The task of answering his official correspondence at the Smithsonian must have been enormous, but he insisted that his staff should follow the example he set in giving every correspondent a courteous painstaking answer. For many years he wrote letters in his own handwriting. Not until the later crowded years were letters written

[18] *Ibid.*, p. 41.
[19] Biographical notice. *Memorial*, p. 469.

in the correct hand of the professional scrivener for Henry's signature.

When a visitor or correspondent failed to respond to his patient courtesy, he developed a line of defense. He would request a proof of their assertions which would take the form of predicting new and important phenomena, the existence of which could be submitted to test either by experiment or observation. "It is not enough that the new system explain facts that we know, for this would be merely exhibiting old knowledge under a new form, but it should point out the way of deducing new facts which have hitherto escaped the eye of the observer or the scrutiny of the inventor."

These tactics proved effective in disengaging him from all but the most persistent fanatics. However, he had to complain to the Regents that there were some dangerous persons whose wiles he could not circumvent.

> Two persons of this class have recently made a special journey to Washington, from distant parts of the country, to demand justice from the Institution in the way of recognition of their claims to discoveries in science of great importance to humanity; and each of them has made an appeal to his representative in Congress to aid him in compelling the Institution to acknowledge the merits of his speculations. Providence vindicates in such cases the equality of its justice in giving to such persons an undue share of self-esteem and an exaltation of confidence in themselves, which in a great degree compensate for what they conceive to be the want of just appreciation by the public. However, unless they are men of great benevolence of disposition, who can look with pity on what they deem the ignorance and prejudice of leaders of science, they are apt to indulge in a bitterness of denunciation which might be injurious to the reputation to the Institution, were their effects not neutralized by the extravagance of the assertions themselves.[20]

Since the secretary's annual report was distributed among the members of the Senate and the House of Representatives, Henry was hereby giving warning to the legislators to be on their guard against the spurious scientists, and offered clues for recognizing the species when its members presented themselves in the guise of injured pa-

[20] *Ibid.*

triots. He was also notifying the legislators that the Institution was not to be used as an agency for advancing the claims of their political supporters. His object was to minimize the mischief that might be done at a time when those with money appeared willing to entrust it to anyone with a pretended scientific background.

The fact that anyone should go to the trouble of carrying complaints to members of Congress that the Institution was withholding recognition indicates the high esteem in which the Smithsonian had come to be held.

An encounter of a different nature related to his conflict with occultism. He did not adopt a supercilious attitude toward the superstitions of the common people. He probably learned many while a boy in Galway, but intimacy with them, while it may have made him tolerant, did not make him sympathetic. His curiosity in regard to them was confined to submitting them to investigation so that a reasonable explanation could be offered for the supposedly supernatural occurrences. There is a long letter written to his wife from Cambridge (August 12, 1859) which contains some interesting information on scientific phenomena and some comment on other matters that were not so scientific.

He told Harriet that he believed the myths concerning mermaids originated with seals. He proceeded to recount the romance of Mrs. Childs, with whom he had been invited to lodge while at Springfield. The invitation was tempting because Mrs. Childs and her daughter had volunteered to show him a jack o' lantern. This recalls to his mind a dispute which almost developed into a quarrel between Torrey and Dr. Jacob Green about a similar phenomenon at Princeton. Henry was curious to investigate, knowing nothing about the properties of methane. The letter concludes with an explanation of the use to which the yolk of an egg is placed during the embryonic state of the chick, and how the head of a ruminating animal is kept in place.

A week later he wrote again to his wife and expressed disappointment that unfavorable weather had prevented him from seeing the will o' the wisp, and after weighing the testimony of those who claimed to have seen it, he was inclined to doubt its authenticity. Having neither information about mermaids nor chicks to impart, he closed this letter with an account of a knife fight and the conduct of the court in receiving the witnesses' testimony.

But Henry was not so tolerant toward spiritualists, to whom he was introduced by Lincoln. The President was no believer in the new craze, but the war had raised emotions to such a pitch that many became believers in the supernatural. Its influence reached the White House through Mrs. Lincoln, who became the victim of one of the imposters who claimed to receive messages from the dead. She eagerly sought the solace of anyone who could communicate with her son Willie, who had recently died.

Mrs. Lincoln enlisted the aid of a medium who had adopted the name of Colchester. This young rascal became a privileged visitor at the White House and succeeded in making an impression upon the President with some of his tricks. Although Henry shrank from contact with these impostors, of whose deceit he was convinced in advance, he could not refuse Lincoln's request to receive the medium and witness a demonstration of his powers. A meeting was arranged in Henry's office in the Smithsonian. Here the medium tried to impress the scientist by creating various sounds which he declared sprang from different indicated corners of the room.

Henry did not respond sympathetically. After having trained his ear to detect and to measure sounds in several score of experiments while testing the acoustic properties of rooms, he was not an easy man to deceive in this respect. However, he gave the medium a patient hearing, obediently turning his head this way and that to listen when sounds were alleged to originate in certain parts of the room. When he had satisfied himself upon the location of the sounds, he brought the private seance to an end by informing the medium: "I do not know how you make these sounds, but this I perceive very clearly—they do not come from the room but from your person."

The medium loftily protested his innocence, but he failed to convince Henry. Doubtless the visitor departed to spread another tale among his followers of scientific prejudice, and so it might have continued to appear had not Henry learned by curious chance how the impostor had produced the sounds.

He was sitting in a railroad coach by a young man who recognized his distinguished fellow-passenger. When they entered into conversation, the young man proudly informed Henry he was a maker of electrical instruments. This aroused Henry's curiosity. He became interested in the other man's occupation and was pleased to receive information upon some of the latest types of electrical instruments. On inquiring the purpose to which some of these instruments were applied, he was surprised to learn they were employed by spiritualistic mediums in the performance of their supernatural tricks. Henry asked the names of some of his clients and was not at all astonished to hear the name of his former visitor at the Smithsonian.

He then described his encounter with Colchester, and confessed his inability to determine how the sounds had been produced, but affirmed his conviction that the medium had created them on his person. The instrument maker assured him he was correct, and he went on to describe the very instrument he had made to enable the medium to perpetrate his fraud. It was fastened around the biceps of the medium's arm, and was so arranged that the clicking or tapping sound was produced at will by expanding and contracting the muscle without moving the arm.

Henry was satisfied to know that his judgement had not erred, and it is probable that he did not hesitate to make known how the deception was performed for he apparently had a bad reputation among spiritualists. In 1868 he expressed a wish to test the claims of a man called Foster, but the man said plainly he did not want Professor Henry to attend his seances. This prompted an acquaintance, Sam Ward, to write to Henry explaining that it would be as well if the latter concealed his identity at one of Foster's seances which Henry proposed to attend. Ward was evidently much impressed by the medium's powers, for he requested Henry's close attention, saying that Foster was capable of wonderful things. The seance proved a disappointment. Henry scribbled a comment on Ward's letter "—a very expert Ingler (?) who exhibited nothing to me of a supernatural character."

Henry was impatient of obscurantism in any form, but especially when displayed by intelligent people. One of his acquaintances was describing his experience at a spiritualistic seance in which, he said, the medium was wafted clear out of the window before his eyes. This was too much for Henry, who burst out indignantly: "Judge, you never saw that, and if you think you did, you are in a serious mental condition and need the utmost care of your family and physician!"

Not all of Henry's visitors and correspondents were impostors, charlatans, cranks, and opinionated inventors. There were many others of a totally different character. The fire which destroyed his correspondence doubtless consumed many papers that would have revealed the sacrament and substance of his thoughts, for it is only when he is dealing with his equals that we can gauge the depth and extent of his mind. The exchange of confidences with the older generation of scholars is lost. The generous acknowledgement of younger men who remembered to recall the kindness with which Henry treated them refreshes and enlivens the arid picture of his benevolent nature presented by the authors of memoirs who wished to celebrate the dead.

Henry had not forgotten those lean years in Albany, and their remembrance held him ever ready to extend a hand to those who were stumbling up the steep ascent of life.

For example, when Simon Newcomb first experienced the urge for authorship to gain a reputation, he wrote out a new demonstration of the binomial theory. He had some doubts concerning the value of his proofs, but he sent it with a note to Henry asking whether it was suitable for publication by the Smithsonian. Henry promptly replied in the negative, but offered to submit the paper to expert mathematicians to obtain opinions upon the merit of the methods employed if that would benefit the young author. This was exactly what Newcomb desired.

This simple courtesy prompted the younger man to visit Henry to express his thanks. He left an amusing account of the visit, which was something of an ordeal. After recording a satisfactory encounter with Rhees, the chief clerk, he went on to say:

> Up to this time I think I had never looked upon a real live professor, certainly not upon one of eminence in the scientific world. I wondered whether there was any possibility of my making the acquaintance of so great a man as Professor Henry. Sometime previous a little incident had occurred which caused me some uneasiness on the subject. I had started out very early on a visit to Washington, or possibly I had stayed there all night. At any rate I had reached the Smithsonian Building quite early, opened the main door, stepped cautiously into the vestibule and looked around. Here I was met by a short, stout, and exceedingly gruff sort of man, who looked upon my entrance with evident displeasure. He said scarcely a word, but motioned me out of the door, and showed me a paper or something in the entrance

which intimated that the Institution would be open at nine o'clock. It was some three minutes before that hour, so I was an intruder. The man looked so respectable and commanding in appearance that I wondered if he could be Professor Henry, yet sincerely hoped he was not. I afterwards found out it was only "Old Peake," the janitor.[21] When I found the real Professor Henry he received me with characteristic urbanity, told me something of his own studies, and suggested that I might find something to do in the Coast Survey, but took no further steps at that time.[22]

Later, When Newcomb applied for a post with the *Nautical Almanac,* Henry gave him a letter of recommendation, having perceived in him a young man of unusual talents. This was Newcomb's initiation to a career in which he earned a well-merited distinction. The younger man never forgot Henry's kindness. Throughout his life he retained the warmest feelings toward the man who had encouraged him in his studies and aided him in securing congenial employment.

Men like Newcomb who fulfilled the promise Henry perceived in their early efforts usually acknowledged their indebtedness. Countless others experienced the same courteous and friendly aid without leaving any record of their debt. Among the notable exceptions was Emile Berliner, inventor of the microphone, the flat phonograph disc, and other useful devices.

Berliner was introduced to Henry by S. S. Solomons, a prominent bookseller who was chairman of the committee appointed to supervise the Morse memorial services. Berliner required advice upon the making of sound recordings upon flat discs, a project which culminated in the production of the now familiar phonograph records. Knowing that no one was better qualified to give advice, Solomons sent him to Henry.[23]

The reader will recall that Henry had employed a crude phonometer to measure the intensity of fog signals. He showed Berliner that his interest in these instruments had not been abandoned, for he pro-

---

[21] He must have been a rather unusual janitor since he was the author of a *Guide* to the Smithsonian Institution.
[22] Newcomb. *Reminiscences,* p. 57.
[23] An account of their interview obviously published by Berliner for publicity reasons appeared in the *National Republican,* October 2, 1877.

duced and demonstrated a Scott Phonautograph, the ancestor of the phonograph.[24] This consisted of a horizontal cylinder mounted on a screw and turned by hand, which gave the cylinder a slowly progressive motion. The cylinder was covered by paper smoked over a sooty flame until it had an even black film. At right angles to the cylinder was a barrel shaped horn, which was closed with a diaphragm, to the center of which was fixed a flexible bristle, so adjusted that its tip just touched the surface of the blackened paper. When the cylinder was turned any sound received into the horn traced a sound wave on the sooty surface. The Phonautograph was probably the first machine to utilize a diaphragm for visual sound recording.

Berliner captured Henry's interest by telling him that he had advanced beyond this, for not only was he achieving the same result upon a flat disc covered with printer's ink, but he was able to reproduce the recorded sounds. He demonstrated other "wonders" of the time, a contact telephone, an electric spark telephone, and a telephonic transfer. The enthusiasm Henry displayed was proof that he had not lost interest in improved apparatus and modes of electrical communication.

The inventor was to go a great deal further as the years advanced, and at the peak of his success he expressed his gratitude for Henry's help and laid his tribute at the feet of the aged scientist who had encouraged his early efforts.[25]

But none exceeded the gratitude of Alexander Graham Bell.

On the bleak first of March, 1873, the elderly Henry had a severe cold that should have been nursed, but he had left his rooms in the East Wing and was at his desk. A visitor's card was brought to him bearing the simple legend "A. Graham Bell, Boston." No man who had travelled so far and had the hardihood to be about his business on that frigid day could be bent upon frivolous business, so the old man did not plead indisposition. He asked that the visitor be shown into his office.

Bell was then a tall slender young man (he was born in the year Henry wrote his first report to the Regents) with black side-whiskers and a drooping moustache. He made an immediately good impression by the deference he showed the older man and by the refinement of his manner.

---

[24] The identical instrument is now preserved in the National Museum.
[25] Wile, F. W. *Emile Berliner: maker of the microphone.* Indianapolis, 1926, p. 100.

He began to talk straightway about an idea he had for the transmission of the human voice over a wire. The chief uses of electricity in those days were for the telegraph and for fire alarms, but inventors galore were harassing the owners of small machine shops with electrical notions they wished to have realized. Henry knew the type well and, had his old heart been a little tougher, he might have been a shade cynical with young Bell. But he was always seeking for some good in the most unpromising situation, and this reconciled him to the young man who had come from Boston to present his idea.

As Bell began to explain the idea, Henry interrupted to ask whether he had seen the latest development along that line. He admitted he had not. Henry rose, and trotted out into another room, from which he presently returned with the instrument.

Charles Boursel, a Frenchman, had written an article suggesting how a voice might be transmitted over a wire,[26] and the German Philipp Reis had taken up the practical problem. He made a telephone which was exhibited at Cooper Union in 1869. It was not a good telephone because it wore out after a short use. Still, it bore the germ capable of improvement, and Henry had bought one to investigate its potentialities.

Bell was interested in the Reis instrument, but he had other ideas. He began again his explanation. He talked well and, as he talked, his eyes glistened with enthusiasm, he gesticulated freely, he ran his hand through his mop of hair.

If Henry noticed the comic appearance of his visitor, he displayed no trace of amusement. He knew when a man talked intelligently about the uses of electricity and when he encountered such a man he was prepared to make allowances for any eccentricity.

When Bell said, rather casually, that "on passing an intermittent current of electricity through an empty helix of insulated copper wire, a noise could be detected" from the nearby magnetic diaphragm, Henry was intensely interested.

"Is that true?" he asked, starting to his feet in readiness to have the fact demonstrated. But Bell was unprepared. He had no apparatus at hand. He offered to give Henry a demonstration at his convenience.

"At once!" announced Henry, all eagerness to become acquainted with a phenomenon he had overlooked.

Bell was all consideration for the older man. He alluded to Henry's

[26] *L'Illustration.* Aug. 6, 1854.

cold and to the inclement weather. Could he not bring his apparatus around to the Smithsonian on the following day, say at noon? To this proposal Henry gave his assent but, before parting from his visitor, he asked:

"Will you allow me, Mr. Bell, to report your experiments and to publish them to the world? Of course, I shall give you full credit for the discovery."

Bell expressed his delight at the offer of this high sponsorship, which was a great deal more than he had anticipated. He departed walking on air.

Next day, the two men, one prominent in the scientific world, the other an obscure teacher of vocal physiology, merged their enthusiasms over a new idea. Henry sat long at the table, listening to the unmistakable sound issuing from the apparatus as the current was passed through it. He was fascinated.

Young Bell was so encouraged by Henry's interest that he ventured to ask for advice on pursuing the project of transmitting the sounds of the human voice "by telegraph." Henry asked him to explain how far he had gone and listened for another half-hour to a description of past and projected experiments.

"What would you advise me to do? Shall I publish the idea and let others work it out, or attempt to work it out myself?" Bell asked in conclusion.

Perhaps Henry recalled the time when he had published an account of his discoveries and had paved the way for others to make fortunes from the telegraph. It was a grave responsibility to advise this earnest young man to forego the reward for his idea. He may have been guided by stern principles in the question of making money for himself, but it was a different matter to impose this guidance upon a stranger.

"You have the germ of a great invention. Work at it," he advised, finally.

Bell was under no illusion. He knew something of the obstacles he must overcome, and the greatest of these was his own technical shortcoming. He mentioned his want of an extensive knowledge of electricity.

"Get it!" was the laconic recommendation.

Bell was not one of those who invited advice in order to do as he pleased. He would not ignore the scientist's advice if it did not happen to coincide with his own prejudgement. He had sought the advice of the highest authority and, having obtained it, was grateful enough to act upon it without delay. That night the elated young man wrote a letter to his parents, telling of the courtesy with which he had been received. He told of the recommendation to acquire a thorough knowledge of electricity. "I cannot tell you how much those two words have encouraged me," he declared.[27]

The circumstances of that meeting were never obliterated from Bell's mind. Henry's consideration for an unknown young man became an example Bell never ceased to follow. At the height of his success he never declined to receive an inventor nor refrained from listening to his story with patience.

"I do not want to discourage him," Bell explained to a friend who remonstrated for the waste of time devoted to one interview. "There may be something in what he has to say. But for Joseph Henry I would never have gone ahead with the telephone." [28]

That first meeting did not end their relations. Henry gave the first public demonstration of the telephone at one of the Saturday night meetings of the Washington Philosophical Society, and it was through his intervention that the instrument received the publicity which brought it into prominence.

Bell exhibited his telephone at the Centennial Exposition at Philadelphia in 1876. The booth he acquired was set up in a small hole in the wall where his exhibit was completely overlooked. The judges gave it no more than a passing glance; wide-eyed visitors were too preoccupied with the great Corliss engine and other examples of giantism to give attention to a gadget hung on a wall.

Then came the day when Dom Pedro, Emperor of Brazil, visited the Exposition and was escorted among the exhibits by the principal officials. Among the group were Sir William Thomson and Joseph Henry. The party would have passed by Bell's little hole in the wall had not Henry chanced to spy the slender figure of Alexander Bell. He steered the Emperor toward the booth and, at Henry's invitation, Dom Pedro placed the receiver at his ear, while Henry spoke a few words into the transmitter. The whole party were electrified when the Emperor exclaimed in his astonishment, "My God! It talks!"

---

[27] Mackenzie, C. *Alexander Graham Bell: the man who contracted space.* Boston, 1928, p. 10.
[28] *Ibid.,* p. 11.

Those four words launched the telephone into popularity. Everyone wanted to see and hear the marvel which had provoked the exclamation.

Thomson and Henry then submitted the instrument to a critical examination. An anonymous commentator says of the incident: "It was a lovely sight when Professor Henry, father of the system, and Sir William Thomson, the greatest living electrician in Europe, met and experimented with the mysterious telephone. Their pleasure reminded me more than anything else of the exuberant joy of childhood, when some beautiful revelation of nature has been brought for the first time to the brain, and when the innocent child expresses happiness in every feature of his face and every movement of his person."

The two scientists had a more practical method of showing their appreciation for Bell's achievement. Both men were on the board of judges and, when they found that Bell's telephone had been omitted from the exhibits receiving awards, they immediately recommended that it be given a Certificate of Merit.

The aged Henry was still keen on experimentation and exploration of new territory. He was no less interested in seeing that honors went to those who had earned them.

At the spring session of the National Academy of Sciences, in April 1878, which proved to be the last over which Henry was to preside, it occurred to Professor Barker of the University of Pennsylvania that it would be a pleasant thing to afford the old president the opportunity of actually carrying out a conversation with some distant point by telephone. The suggestion was accordingly addressed to William Orton, president of the Western Union Telegraph Company, who answered as follows:

> I note what you say concerning the meeting of the Academy of Sciences at the Smithsonian Institute on Tuesday, and I sympathize with you most warmly in your desire to exhibit to Professor Henry the latest wonders of that science to which he has devoted so much of his priceless life, and for which he has received so little reward. But if the world is slow, as it often seems to be, in doing justice to those who have done most in promoting the interests of science, and thereby the welfare of mankind, I believe that justice is sure to be done in the end, and so believe that the time will come when all men, everywhere, will recognize his

services in connection with the grand results of his noble life. Any service that I can render toward making the occasion to which you refer interesting to all who participate in it will be rendered most cheerfully.[29]

The local offices of the telegraph company were instructed to place their wires at the disposal of the scientists for the projected experiment, and on the afternoon of April 18 the line between Washington and Philadelphia was connected with a telephone in Henry's room in the Smithsonian. As soon as the connection was made, a well-known voice, that of Mr. Bentley of Philadelphia, was heard reciting one of those classic nursery rhymes dear to children and for which learned and reverend professors sometimes retain a lingering fondness.

Henry then told Mr. Bentley over the telephone of the pleasure and gratification he felt, and which by his manner he showed to those who were present. "Doubtless," said Professor Barker, who had been the prime mover in the experiment, "his memory carried him back to 1830, when in the laboratory of the Albany Academy he made the early researches, of which the interesting experiment he now assisted at, and which showed that men could converse through one hundred and forty miles of intervening distance, were the outcome."[30]

[29] Quoted in Pope, F. L. Life and work of Joseph Henry. *Jour. of Amer. Elect. Assoc.*, 1878. Vol. II, p. 134.
[30] *Ibid.*

CONSUMMATION

When the Civil War had ended, Henry's major works were done. The nation had undergone many changes in the first three-quarters of the nineteenth century. Most of these changes had their source in advances of knowledge and Henry was alert to observe how each addition to knowledge might be utilized to advance some other phase. He stood in his declining years like a signpost pointing out the way for others.

The task of promoting knowledge had become his life's labor, and he never slackened his efforts. Alike in his practical work for the government bureaus, in his immediate labors for the Smithsonian Institution, and in his encouragement of other bodies and individuals, the impress of his kindly nature and his unmistakable ability were felt.

He never forgot that through science he had gained what position he held in the world, nor did he relax his efforts to plead for the recognition of the scientist's right to social recognition. In one of his early annual reports he had introduced this argument:

> While we rejoice that in our country above all others so much attention is paid to the diffusion of knowledge, truth compels me to say that comparatively little encouragement is given to its increase. . . . As soon as any branch of science can be brought to bear on the necessities, conveniences, or luxuries of life, it meets with encouragement and reward. Not so with the discovery of the incipient principles of science; the investigations which lead to these receive no fostering care from the government and are considered by the superficial observer as trifles unworthy the

attention of those who place the supreme good in that which immediately administers to the physical needs and luxuries of life. But he who loves truth for its own sake, feels that its highest aims are lowered and its moral influences marred by being continually summoned to the bar of immediate and palpable utility.[1]

These annual reports to the Regents provide a running commentary upon the progress of science, and exhibit proof of the catholicity of Henry's interests as well as the penetration of his mind and the strength of his convictions. The correspondence he selected for publication is evidence of the weight attached to his judgement and advice (and that of the entire Smithsonian staff) by men in all parts of the world. Opinions were sought on innumerable subjects.

As Henry aged he grew in honor. The great institution to which he had pledged his life assumed the character he had designed for it. Its purpose had been warped almost beyond recognition by well-meaning but misguided enthusiasts who had been influenced by popular demands. Henry incurred no small amount of criticism in his efforts to attain the object he had in mind, but by sheer force of character and singleness of mind, he removed the undesirable elements.

Some projects that were attached to the Smithsonian, Henry thought, had no place in the provisions of the bequest, and these he labored to remove outright. Others, such as the organization for reporting weather, were nourished until they could stand on their own feet or be taken over by some other more appropriate agency. Thus the Institution was left free to take on new projects.

Henry lived to see most of the provisions he had made for the Smithsonian reach a happy fulfillment. In 1866 he witnessed the transfer of the odd assortment of books in the library to the Library of Congress. When one surveys the extent of that superb collection today one cannot refrain from wondering what would have been the outcome of rivalry between it and the Smithsonian library had the original provisions for the latter been carried out.

No reproach could be levelled against Henry for the manner in which he had administered the growing museum collections. His efforts to secure a good collection of botanical specimens through Torrey, his interest in acquiring historical material evident in his acceptance of Hare's equipment, and similar actions, had proved his

[1] *Smithsonian Annual Report for 1853*, p. 8.

willingness to conform within limits to the official program of the Institution. Although the greater burden of this work fell upon the faithful Baird, it was Henry's duty to supervise the distribution of more than one hundred thousand specimens, correctly identified and labelled. The work associated with the collections of natural history specimens made by the various surveying parties was becoming a very severe strain upon finances and, while Henry never disputed the wisdom of making these collections, he was unceasing in his efforts to divest the Smithsonian of the responsibility for their maintenance. He looked with dread upon any further acquisitions as an additional encumbrance.

He had the foresight to perceive that the projected Centennial Exposition to be held in Philadelphia in 1876 would have a decisive influence upon the future of the National Museum he was advocating. He foresaw that the great collections assembled by government departments for display at the exhibition, and those specimens presented by foreign countries, would ultimately find their way to the Smithsonian, where there was no accommodation for them and where their maintenance would be a further burden upon the inadequate income. At his suggestion the Regents asked for an appropriation to provide a special building to house the National Museum. The necessary bill was passed in 1879.

The National Museum, founded after his death, none the less bears the impress of Henry's conception of its functions. It serves as a major reference museum in natural science. No interpretative account of its material in terms of political events or economic consequences is sought. The museum aims to enrich the knowledge of those who already possess the key to the proper continuity of events; for the less well informed it serves as a shrine where historic relics may be seen and reverenced.

The pictures were transferred to the Corcoran Gallery when that collection was willed to the nation, and Henry was appointed to the Board of Trustees.

Thus the policy originally outlined in the handling of the meteorological service was generally followed. As soon as this service had been well established upon sound principles, it was transferred to a government department where its usefulness was apparent and its continuity assured. The Department of Agriculture's comprehensive study of forest trees originated with Henry, and was conducted under his supervision until the project grew beyond the capacity and resources of his staff. In measuring Henry's achievement we cannot afford to overlook the fact he was the moving spirit in the creation of the National Zoological Park, the National Museum, the Bureau of Ethnology, and the National Gallery of Art.

It was astonishing how wisely he had administered the Institution's finances while it was saddled with these unwanted departments. Of necessity, Henry had to exercise the strictest economy in administering the funds in his charge, but it might be argued he carried it to excess. He imposed upon himself an unnecessary burden. No expenditure could be made without his personal approval. Not so much as a whiskbroom could be acquired without a full explanation of the need for it. The only justification for saddling himself with these trifling details was the fruit borne as reflected in the balance sheet. Long after Baird had proved himself a capable administrator he had to submit all his proposed expenditures for Henry's approval, and it was only in the final year when the secretary had to unburden himself of some responsibility that the head of the museum was permitted to take charge of the appropriation for his department.

Henry's entire life had been devoted to work for the benefit of mankind. His vacations were still allotted to gratuitous labor for the Light House Board and on other government problems. Finally it was realized that unless he was specifically directed to go away and rest, he never would do so of his own volition. In 1870 the Board of Regents granted him three to six months leave of absence to travel abroad, and to make sure that he would raise no objections upon the score of expense, they declared the journey would be undertaken on behalf of the Institution, and he was given $2,000 for expenses.

Henry was completely taken by surprise when this scheme was launched for his benefit, but he was overcome with joy when he recognized it was merely the symbol of the affectionate regard in which he was held by the members of the Board. Perhaps he carried his head a little higher when the representatives of the Cunard and Bremen Lines of steamships began to compete for the honor of providing free accommodation for a passenger who had done so much to promote the mariner's safety.

He sailed with his daughter Mary as companion on June 1, 1870, and was absent for four and a half months. The outbreak of the Franco-Prussian War interfered with his plans and restricted his movements, but he contrived to visit England, Scotland, Ireland, Belgium, Switzerland, and at least to set foot on French and German soil.

The Europe which he now saw had changed considerably during his twenty-four years' absence. A series of fruitful investigations touching "men's business" through the practical consequences, and "men's bosoms" through their revolutionary effects upon thought, had wrought notable changes in all the nations.

At the time of his first visit men were concerned with empirical inventions like the steam engine, but science was now teaching man where and how to look for the sources of power he sought to control. Men everywhere were learning that they lived in a universe governed by fixed laws which could be discovered and turned to good account. Forethought and exact knowledge were replacing rule-of-thumb.

Germany was in the flood-tide of her progress, and had organized her educational system in order to prepare an educated people. England had fallen behind. She was markedly inferior to Germany in education, in most respects behind France, and in a few vital aspects had lost ground to America. She had an abundance of schools but no educational system. Royal Commissions had been appointed to seek remedies for this defect and, as one was now conducting an inquiry into conditions of teaching science in public schools and universities, its members could not forego the opportunity of consulting the visitor who had done so much to promote scientific thought in America.

This meeting brought Henry into touch with some of his peers of a new generation, successors to his old friends. He submitted to questioning by Huxley, Lubbock (later Lord Avebury), and G. G. Stokes (secretary to the Royal Society) on the activities of the Smithsonian Institution and upon the extent to which science was being taught in the colleges and schools of the United States. Although he was not able to furnish all the information asked for by the educational experts, Henry acquitted himself with credit, displaying a surprising amount of knowledge regarding science education. His views were decidedly "liberal" for, after the Darwinian controversy, science teaching (when it did not encounter bitter religious hostility) was suspected of being flavored with radicalism.

He must have been gratified when the members of the Royal Commission, expressing their admiration of the work accomplished by the Smithsonian, expressed the opinion that the success it had achieved was dependent upon the unusual combination of character and ability to be found in the man who directed its activities. They said that any similar institution must first find a man of the same breed before it could hope to rival the Smithsonian's success and influence.

The beauty of the world had a strong appeal for Henry, and much of the time spent on this journey was devoted to visiting the scenes which had inspired his favorite descriptive poetry. He loved the sweet peace of the countryside, perhaps because it was a landscape upon which humanity had set its mark. But scenes of wild natural beauty touched a deeper chord in his emotions. Mountains have held a fascination over many scientists. Faraday had fallen under their spell. Arnold Guyot had spent his youth in the Bernese Oberland, in view of the Jungfrau and the Schreckhorn. Bache had visited this neighborhood in 1837 and had written and spoken of its grandeur in enthusiastic terms. Agassiz had told Henry of the abundant joys a scientist might find in the Oberland, the most graceful and romantic of the Alpine masses, and it was upon his recommendation that Henry had included this region in his itinerary. It is the home of long glacier-passes leading through great varieties of mountain scenery. Somewhere about here, Agassiz had lived under a slab of gneiss while studying the Aar glacier. Henry went in search of the spot.

He passed from the homely beauty of the lower slopes, the thickets, the falling streams, the flowers, and climbed toward the grim dark peaks. As the party crossed a shoulder of the mountain, they came within sight of the glacier lying at the foot of the Finsteraarhorn. There he felt closer to the inner being of his friend. The party returned and, having crossed the mountains, descended into the Rhone Valley. At a sudden turn in the road they came within sight of the majestic Rhone Glacier.

It was a sight of indescribable magnificence. In front lay the mass of the Finsteraarhorn; further back was the grand peak of the Weisshorn (first conquered by his friend Tyndall); to the left lay the snowy summits of the Monte Rosa, by the side of which rose the black cone of the Matterhorn, streaked with snow.

The sublimity of the view of the mountains, cold and unchanged, with their frozen tranquility and their indomitable strength, held Henry speechless. He stood in awe before their splendor.

Finally he said in a voice choked with emotion, "We should go no further. This is the place to die!"

CONSUMMATION

To the man whose natural earth is the pavement or the concrete highway, this may not be apprehended, neither can it be explained. But there is nothing remarkable in this expression of sympathy for nature. After all, it was the medium in which he had worked all his life.

Although he knew and loved so much in inanimate nature, yet he retained a large opinion of human nature. One of the rewards of this journey was to establish personal association with men whose work was familiar. Most of those whom he had met on his first visit to England were now dead, and a new generation had stepped into their places. These new men he knew through their writings and letters. To meet a good man through the medium of cold print is only to whet the appetite for closer acquaintance, and the understanding of the inner sanctuary of a friend's mind cannot be complete until one has sat in an armchair and exchanged views with him across the hearth, or walked through a country lane in his company.

While in London, Henry met John Tyndall, Faraday's successor as Resident Professor at the Royal Institution and adviser to Trinity House. Henry knew Tyndall's work through their common interest in lighthouses and fog signals. They instantly formed a friendship which was to have a remarkable memorial. When Tyndall came to the United States in 1872–1873 to give a series of public lectures on the subject of light, Henry was glad to repay his hospitality. After deducting his personal expenses for the journey, Tyndall placed the remainder of the fees he had earned in Henry's trust, to be devoted to some educational project. Henry decided that the interest from the fund should be used in providing scholarships at Harvard, Columbia, and Pennsylvania Universities.

The prolongation of the journey to Scotland was mainly to meet Sir William Thomson, the future Lord Kelvin, who was bringing fame to Glasgow University. Friendships established with men like Thomson were charged with precious reminiscences. Soon he was to have the pleasure of acting as host to these brilliant Britons, and to introduce them to the people of the United States.

On his return Henry was asked by the Regents to present them with a report upon his journey, but he never did so. His inability to write and to speak of his own doings, unless duty compelled him to do so, caused him to neglect the preparation of a record that would have been an interesting chapter in his life. Or perhaps he found himself immersed in a multitude of duties and interests of more pressing importance, and never found time to do any superfluous writing.

The Institution still had first call upon his time; correspondence consumed several hours of his day. There were details still to be arranged concerning the divestment of the museum and the library. Soon he was to be caught up in the mass of organization involved in the Centennial Exposition.

He still regularly attended the meetings of the National Academy, the American Association for the Advancement of Science, and his Saturday nights were devoted to the meetings of the Washington Philosophical Society, which had recently been formed. His lively comments upon the papers presented at the meetings showed that his mind had gained in breadth.

But he was an old man and this fact was borne upon him by the passing of old friends. Silliman had died in 1864, Bache in 1867, Agassiz, Joseph Saxton, and Torrey in 1873. But while a younger generation were arising to take their places in the world of science, none could quite replace the old friends who had shared the days of struggle. He had arrived at a period which is often marked by discouragement and disillusion. Henry was free from this malady of oncoming age. He retained a surprising freshness of outlook, possibly through revivification by the younger men who sought his counsel. His principal confidant now was Simon Newcomb.

There was still flexibility in his thought; there was no stiffening of the mental joints. The writing in his later annual reports exhibits nothing of a doleful tired old man who has passed through the ordeal of life's normal span. These writings have the crispness and freshness of his Princeton days.

It should be evident that if Henry had done nothing but attend to the work incidental to the Smithsonian and his societies he would have been busy enough for a man of his age. But when the heat of summer came around, Henry allowed himself no repose. He applied himself conscientiously to the unfinished work of the Light House Board. There were experiments and tests to be made in the little laboratory on Staten Island. A portion of the time was spent pleasantly on the boats and steamers of the Board, but it was hard work, nevertheless.

Only at night would he lay aside his work. Then would he yield to the temptation of watching the processional planets slowly march and wheel on their dignified courses. Or he might watch the lightning flicker across the sky.

During the summer of 1877 he was at work in the laboratory. The tail-end of the season turned out to be chilly, blustery, and wet. Weather was not permitted to interfere with the program of work he had planned. He went on working until the summer passed and autumn was merging into winter.

"After an almost uninterrupted period of excellent health for fifty years," he said in the opening address to the National Academy of Sciences,[2] "I awoke on the 5th December, in my office at the Light-House Depot on Staten Island, finding my right hand in a paralytic condition. This was referred at first by the medical adviser to an affection of the brain, but as the paralysis subsided in a considerable degree in the course of two days, this conclusion was doubted."

On returning to Washington he placed himself in the hands of family physicians, Drs. N. S. Lincoln and G. Tyler, and under their ministrations the paroxysms of pain through the region of the heart, from which he also suffered, subsided. He was restored in some measure to his customary health. Reluctantly he consented to relax his daily routine, and allowed his able assistant, Professor Spencer F. Baird, to carry a fuller share of the work. Through this division of labor and the care of the physicians he confidently expected to recover his old energy.

However, he did offer his resignation as president of the National Academy, but he was persuaded to withdraw it and to allow the vice president to relieve him of any duties he found burdensome.

But Henry did not make the complete recovery he had hoped for. It was decided he should consult Dr. S. Weir Mitchell, America's first neurologist and a distinguished novelist. Mitchell made a thorough examination and diagnosed kidney trouble. He advised a strict curtailment of work and suggested that Henry remain in Philadelphia for special treatment.

"Am I mortally ill?" asked Henry bluntly.

"Yes," answered Mitchell, knowing his man and realizing the folly of equivocation. Nephritis, or Bright's disease did not, in those days, permit much hope for recovery.

"How long do you give me to live?" was the next question. When the physician demurred at fixing the date, Henry persisted: "Six months?"

[2] Read April 16, 1878.

"Hardly that."

For a few more minutes they talked of the arrangements for the treatment Mitchell had prescribed, but neither speaker desired to prolong the conversation. Finally the moment came when Henry had to order his carriage for the doctor and, as Mitchell was about to leave the room, his patient recalled a duty unperformed.

"I have not discharged my material obligation to you. How much do I owe you?"

"You are not in my debt," answered Mitchell. "There are no debts for the dean of American science."

The old man was deeply moved. For a moment he could not find the appropriate words to express his gratitude, and when he found his voice it was to say humbly, "I have always found the world full of kindness to me, and now here it is again."

"You do not remember, sir, that once you said to me, a small boy, when you had been kindly attentive and I had tried to express my thanks, that perhaps the time would come when I could oblige you in return. If this obliges you, my time has come." [3]

Sure now that the days remaining were numbered and that little time was left to complete his work, knowing also that he would have to leave many things undone, yet he uttered no word of complaint or lamentation. He had to curtail still more of his official duties and to abridge his social diversions in order to husband his strength. But otherwise he was not changed by the sentence of death. For his friends, there was the same benign smile, the affable greeting, the warm handclasp they had always known. For strangers there was the unfailing courtesy of former days.

In the outer world there was an awakening consciousness that justice had not been done this old man who had given the brilliance of his mind to the benefit of his fellow men without expectation of reward. A group of his friends familiar with his refusal to compete for pecuniary rewards, and fearful that he had been unable to make provision for a helpless old age or for his beloved wife and daughters, decided that his small salary should be augmented by voluntary gifts to a fund which should be placed at his disposal.

Among them they raised the respectable sum of $40,000, which it was proposed to turn over to him as a symbol of the affection and re-

---

[3] Mitchell, S. W. *Address*. Semi-Centennial Anniversary of Nat. Acad. of Sci. Washington, 1912, p. 89.

spect in which he was held by the scientific world. This gift, it was estimated, would add $2,400 a year to his income.

The presentation was made at the April (1878) meeting of the National Academy of Sciences, at which Henry offered his resignation. He naturally shrank from accepting the gift for his own use but the promoters had a shrewd judgement of his sensitive nature, and had arranged for the payment of the interest to Henry during his lifetime and, after his death, to the surviving members of his family. On the death of the last survivor the fund was to pass to the National Academy to be used for assisting meritorious investigations, especially in the nature of original research. The implication of solicitude for his family, for whom he had been unable to make any suitable provision, made rejection of the gift impossible.

After Henry had been treated with so much honor by his fellow workers and returned through the deserted corridors of the great Institution to his private rooms, he must have experienced a great joy. In renouncing the material gains the world had to offer he had gained the affection of those whose opinions he most respected.

As his days swept all too swiftly to an end, he spoke of approaching death only with those who were admitted to close intimacy. One would have thought he might have spoken freely of religion during these days but beyond the letter to his friend Patterson [4] he does not appear to have given any particular consideration to the subject. He seems to have awaited the end with serenity and quiet confidence.

In the early days of May he received a visit from an old friend who had journeyed from New York to see him. Both were aware that this meeting would be their last. Henry expressed no regrets. "I may die at any minute now. I would like to live long enough to complete some things I have undertaken, but I am content to go. I have had a happy life, and I hope I have been able to do some good."

He had a long last talk with Simon Newcomb, the younger man whom he had guided at the outset of his career. They skirted the inevitable. Instead of talking about his approaching end, Henry listened eagerly to his visitor's account of the transit of Mercury, and they discussed the occurrence. An interest in the ways and truths of the universe could not expire until life itself was done.

[4] See page 296.

One of his admirers,[5] in speaking of this period, recalled the tranquil spirit of Socrates who, as the chill of the fatal hemlock crept over his heart, uncovered his face to ask his disciple Crito not to omit payment of a small debt to Aesculapius. In what was probably his last hour of consciousness, Henry directed the dispatch of a letter of courtesy which, on the day before, he had promised to a total stranger.

The time arrived when he who had so openly rejoiced in life on this earth saw before him the gateway to an ampler world. The end was calm and peaceful. After a life of honest labor, he accepted death without fear. When the last sleep was on him his thoughts were with his incompleted tasks. Those who ministered to his needs heard him speak in his slumber of the acoustical experiments he had been making on the Light House Board's tender *Mistletoe,* and of making notes on electric charges sent through wires. So, at the end, the first and the last tasks of his scientific life lay close to his mind.

He was sleeping quietly when midday struck. Ten minutes later he was dead. The day was May 13, 1878.

Clark Mills, the sculptor, made a cast of his face.[6]

He was buried in Rock Creek Cemetery, Georgetown. President Hayes, the members of the cabinet, the justices of the Supreme Court, members of the diplomatic corps, and representatives of numerous learned societies followed him to the graveside. Congress paid him unusual honors. By a concurrent resolution of both houses, a memorial service was held in the House of Representatives on January 16, 1879. Few men in this country have been the subject of so many formal eulogies. But eulogy was the last thing such a shy and modest man would have sought or desired.

Ten years after Henry's death, Senator Morrill of Vermont proposed a bill appropriating the sum of $15,000 for the erection of a bronze statue to commemorate Henry's contributions to the knowledge of mankind. This statue, the work of William W. Story, which now stands in the Smithsonian grounds, was unveiled in 1883.

[5] Welling, J. C. Life and Character of Joseph Henry. *Bull. of the Phil. Soc. of Washington.* Vol. II, p. 203.
[6] Philadelphia *Public Ledger.* May 17, 1878.

HENRY'S PLACE IN SCIENCE

Much of Henry's tribute to Peltier can be said of Henry himself: "He possessed in an eminent degree the mental characteristics necessary for a successful scientific discoverer; an imagination always active in suggesting hypotheses for the explanation of the phenomena under investigation, and a logical faculty never at fault in deducing consequences from the suggestions best calculated to bring them to the test of experience; an invention ever fertile in devising apparatus and other means by which the test could be applied; and, finally, a moral constitution which sought only the discovery of truth, and could alone be satisfied with its attainment." [1]

But Henry is more than a figure of individual merit. He is a key to the understanding of nineteenth century science in America.

He lived through three-quarters of a century in which the world made greater progress in the application of science to the useful arts of life than in any similar period of its existence. Until the beginning of the nineteenth century chemistry and physics had been ignored by the general public. The study of these subjects did not appear to serve any useful purpose, and the culture of the period was far from taking into account the philosophical value of scientific and mathematical studies. Science was detached from the popular consciousness. The absence of popular understanding undoubtedly aided in driving the scientist into intellectual isolation.

The industrial revolution was in full progress in Europe, but its rapid growth as exemplified by the application of steam power to the coal mines and textile mills had little connection with the groups of

[1] *Smithsonian Report for 1867*, p. 158.

natural philosophers in the universities, although eventually the study of such machines as the steam engine led to scientific advance.

Undoubtedly the dominant figure of scientific thought during Henry's lifetime was Sir Isaac Newton, who had died a century before. Newton's work had led to such fruitful and diverse ramifications that it was accepted as the framework into which all scientific progress had to fit. The Newtonian world-picture, through the philosophies of Locke and others, even carried over into politics, where it may be seen in the Declaration of American Independence, for example, and it has been a continuing influence through Henry's time to the present.

Small wonder, then, that Henry worked and thought in the Newtonian tradition. He was one of those Newtonians described by Roger Cotes in his preface to the second edition of the *Principia:*

> These derive the causes of all things from the most simple principles possible; but, then, they assume nothing as a principle that is not proved by phenomena. They frame no hypotheses, nor receive them into their philosophy otherwise than as questions whose truth may be disputed. They proceed, therefore, in a twofold method, synthetical and analytical. From some select phenomena they deduce by analysis the forces of nature, and the simple laws of forces; and from these by synthesis show the constitution of the rest.

Henry's intuitive conception of electricity and magnetism as imponderable fluids in the framework of Newtonian mechanics may have been a hindrance to his complete final understanding, but they furnished a useful hypothesis for experiment, and Henry was always keenly aware of the tentative nature of such hypotheses. The hypotheses were useful so long as they led him to new experiments and observations of new effects, and there is no question that they succeeded to this extent.

Henry's experiments bear witness to this. He wrote: "Truth, as has been properly said, belongs to mankind in general; our hypotheses belong exclusively to ourselves, and we are frequently more interested in supporting or defending them than in patiently and industriously pursuing the great object of science, namely, the discovery of what *is*." [2]

[2] Comm. of Patents. *Agricultural Report for 1856,* p. 456; *S.W.,* Vol. II, p. 34.

Henry's mind moved reluctantly toward any end that he might not attain by direct apprehension, and he did not always submit new facts to a sufficiently vigorous logical analysis. He had the good fortune to find at hand an aspect of scientific work which gave free play to his talent for experimentation. That he was a first rate experimentalist cannot be denied, and this rare talent was a prime necessity at the dawn of electrical science.

We should not underestimate the function of the experimental scientists, nor should we assume that experimental science leads infallibly to the discovery of new truths. An investigator has to possess a disciplined freedom of thought, and the clarity of vision which enables him to seize the essential and to discard the unessential.

Henry's work belongs to the history of science and deserves to be accorded more prominence than it has received. It is not enough that he should be mentioned in a footnote to Faraday's work on mutual induction. His discovery of self-induction is of first importance. Although his omission to announce the results of his experiments promptly has been the primary cause for Henry's neglect, contemporary conditions were not without their effect. The time had not yet arrived when new ideas were readily assimilated, and the remoteness of the New World from the Old in those days of slow communication ensured a more rapid diffusion of Faraday's achievements through the centers of learning. When the news of Henry's accomplishment reached Europe it bore the appearance of mere confirmation of what had already been done.

Henry lacked the divine spark of the superlative genius. He too rarely abandoned patient research in order to follow a brilliant intuition. Like a child he went on asking "Why?" without seeking a final answer. With the naivety of childhood he kept his freshness of outlook, and this seems to have encouraged him to plunge into new situations with zest. As a result, his scientific life was a series of beginnings.

Experimentalism in physics has the disadvantage of producing diminishing returns of response to effort applied. Mathematics expands the horizon and blazes new trails. In spite of keen mind, quick eye for observation, and an unusual power of grasping the workings of things as they were revealed to a qualitative observer, Henry suffered because of the limitations of his mathematical equipment. He thought in terms of physical bodies obeying mechanical laws, and he did not attempt to express his findings mathematically (see Appendix). He was probably the last physicist to attain international fame by the use of simple arithmetic as his sole mathematical tool. Had he taken pains to become familiar with Sir W. R. Hamilton's work, published in 1834, in which was shown that all gravitational, dynamical, and electrical laws could be represented as minimum problems, he could have added significantly to his discoveries.

This neglect of mathematics is all the more singular because when Henry was Secretary of the Smithsonian Institution he recorded the progress of the more abstruse mathematical theories in astronomy and geodesy in his annual reports. These reports show that he was in touch with the ablest mathematicians in the country. In his "Syllabus of a Course of Lectures on Physics" [3] occurs a pair of paragraphs showing his appreciation of the part played by mathematics in physical studies:

(18) Importance of mathematics in the study of physical science, principally used in the process of deduction.

It is the great instrument of all exact enquiry relative to time, space, order, number, etc. And as the material universe exists in space and consists of measurable parts, and its operations are produced in time and by degrees, the abstract truths of mathematics are applicable by analogy to the developments of those of external nature.

A facility in handling mathematical symbols would hardly have aided him in the discovery of induction, but had he applied the differential calculus to the study of induction he could have carried it far beyond the limitation imposed by the retention of a mechanistic idea of its mode of action. And if, during the time of his greatest activity while at Princeton, he had raised himself over the barrier created by his natural inclination toward experimentation and had plunged boldly into the wider field of mathematical theory, there is a tenable supposition that he could have reached an even higher peak of achievement. But he shrank from the suggestion that forces were possessed of qualities that had no counterpart in mechanics.

An historical research into the accomplishments of Joseph Henry must include two parts. First, there is the proof, based upon printed words, that he accomplished certain things. That is affirmative truth, and can be definite. The second step lies in showing that he had not

[3] *Smithsonian Report for 1856*, p. 191.

been anticipated. Here an assertion is less susceptible to definite proof. However, the student of Henry's work is hampered in the affirmative proof by the fact that records are sometimes missing. While his publications and the testimony of those who witnessed his experiments reveal an exceptional volume of accomplishment, it is probable that the more complete record of his own journals would have added other important contributions had they been preserved.

In the attempt to determine in what cases his discoveries were prior to those of other men we encounter difficulties. The means of rapid communication were entirely lacking one hundred years ago. As a result scattered scientists working in the same fields made similar discoveries without realizing it. Because the original notes of some investigators are not available, the credit for discovery has generally been associated with priority of publication. Even in Europe, with its shorter lines of communication and more voluminous literature, conflicting claims for priority were sometimes advanced, and occasionally the scientific accuracy of the investigators themselves is open to question where matters of priority were concerned.

This book attempts an accurate appraisal of Henry's work, and in a sincere effort to resolve the uncertainties surrounding his accomplishment the work of his contemporaries has been carefully weighed.

The Albany period comprised the only continuous series of physical investigations which anyone up to that time had conducted in America. Henry's powerful electromagnet was probably the greatest single utilitarian electric invention at that time. Had the inventor gone no further he would be remembered for this alone. The discoveries he made in this period are impressive. In summary they are as follows:

1. By using insulated wire and by devising windings of more than one layer, he was able to effect great increases in the strength of electromagnets to the extent that they were of practical use. His form of construction is used universally for electromagnets today (1829–1830).

2. By demonstrating that the coils of the magnets constructed by this method could be made up of either one long wire or several shorter ones, he showed the proper relative proportions of magnet to battery for maximum effect in these two cases. He pointed out that the intensity magnet and battery were suitable for operating over long wires, while the quantity magnet and battery were suitable for local circuits. This was the first experimental demonstration, made independently of Ohm's theory, of the necessity for properly matching the

resistances of the different parts of an electric circuit, a principle which has since become an important requirement in the design of all electric circuits (1830).

3. He was the first to describe and to construct an electric motor using electromagnets and a commutator. This motor was reciprocal rather than rotational, but it contributed fundamental elements of permanent value, and it served to stimulate interest in the problem of deriving mechanical power from electricity (1831).

4. In spite of Barlow's discouraging experience, Henry profited from his knowledge of circuit-matching to make the first electromagnetic telegraph with polarized armature, which made rapid signaling possible, and which has formed the basis of practically all subsequent commercial wire telegraphy, and his receiver gave the first audible signal (1831-1832).

5. He and Faraday independently discovered mutual induction and self-induction. Purely upon the basis of prior publication, the discovery of mutual induction is accorded to Faraday, and the discovery of self-induction to Henry (1832).

Now let us turn to the Princeton period, likewise rich in accomplishment. In recapitulating the achievements of this period we find the outstanding features to have been:

6. The development of the electromagnetic relay, which Henry communicated to Wheatstone in 1837 while visiting England (1835).

7. Non-inductive windings, the earliest reference to which is contained in a verbal communication to the American Philosophical Society in 1835.

8. Both Faraday and Henry used a transformer structure in their first mutual induction experiments but it was later discovered by Henry that by proper proportioning of the coils the voltage could be stepped up or stepped down with this structure. This is the basis of the modern electrical transformer (1838).

9. The relations of high-order induced currents revealed in his paper on "Electro-Dynamic Induction" (1838).

10. Electromagnetic shielding, described in a paper presented to the American Philosophical Society in 1838.

11. Variation of induction between coils with separation, the action of induction at a distance, and the oscillatory nature of a discharge from a Leyden jar, all of which discoveries had an important bearing upon the development of radio telegraphy and telephony (1842).

These contributions by Henry to electrical knowledge have been essential to practically every commercial application of electricity. Consider some of them, even at the risk of repetition.

The idea of the electric telegraph was not new when Henry approached it, but the experience of others tended to show that the development of magnetism in soft iron at a distance was impracticable. In spite of this discouraging attitude, Henry persevered and was able to make three contributions which made the telegraph practicable.

He demonstrated that magnetic action and magnetic control could be exercised at considerable distances if the battery and the magnet windings were suitably proportioned. This is the basis of all electric telegraphs today. He discarded the galvanometer or compass needle as a receiving device and substituted a magnet operating a movable armature, thus making possible rapid signalling and audible reception. This continues to be the basic form of telegraph receiver, even with the modern printing telegraph systems, in which electric typewriters are controlled by magnets of the Henry type. He constructed and operated the first electromagnetic relay, a device by which the current in the line circuit controls an armature which carries the contact of a local circuit so that the feeble line current, instead of directly controlling the receiving mechanism, merely opens and closes a local circuit and a strong local current performs the necessary function in the receiving mechanism.

These three developments constitute the fundamentals of a complete magnetic telegraph system. From this start the telegraph rapidly grew to a world-wide communication system as his successors perfected the details and commercially exploited his discoveries.

The next important commercial application of electricity came some thirty years after Morse established the first commercial telegraph line, when Bell developed the telephone. Bell's telephone made important use of Henry's magnets, as does every telephone receiver today. The telephone bell which attracts the attention of the party called consists essentially of a polarized ringing device suggested by the receiving mechanism employed by Henry in his first demonstration of the telegraph. In all telephone switchboards there are multitudes of electromagnetic relays. The prototype of the millions of relays operating in the telephone systems of the world (not counting millions of others performing other functions) was Henry's electromagnetic relay of 1835.

Telephony owes yet another debt to Henry. He discovered self-induction which must be taken into account in many phases of electrical design. In long telephone lines it can be arranged to produce a favorable effect upon transmission.

The start of commercial electric light and power systems, which came about 1880, was likewise built upon Joseph Henry's work. Powerful electromagnets are the basis of every generator and motor, and it would be difficult to build either without incorporating Henry's ideas. The commutator which first appeared in Henry's motor is an essential part of every direct current generator and motor. The first electric motor is rightfully ascribed to Faraday, but it had no commutator and no electromagnet was employed. Both features appear in Henry's reciprocating motor.

While almost every branch of electrical application is under debt to Henry, the alternating current system now in general use, especially long distance power distribution, is under an especial obligation. In his work during the year 1838 he demonstrated that through the proper proportioning of the windings of two coils in inductive relationship, the voltage in the secondary current could be stepped up or stepped down; and this was the genesis of the transformer. The transformer is fundamental to alternating current systems and without it alternating currents could not have become so generally adopted. Through its use it is possible to design generators for operation at the most effective voltage, and then to step up the voltage by means of a transformer to the most efficient level for use on the transmission lines, and then at the distant end of these lines, by means of other transformers, step down the potential to the most efficient and convenient voltage for use on distributing systems.

Henry also made several contributions to radio communication. His work in detecting the discharge from Leyden jars at several hundred feet clearly foreshadows wireless transmission of messages. His own comment and that of his student William J. Gibson [4] indicates that he had noted the discovery of the germ of radio broadcasting and transoceanic telegraphy and telephony.

The fire which destroyed his records probably deprived us of many other interesting and suggestive things which Henry had done, but it is highly improbable that the preservation of this material would have revealed any other discovery comparable to those enumerated above.

[4] Quoted on p. 142.

Henry lived to witness an extensive application of telegraphy, and he saw the invention of the telephone but not its commercial application on a large scale. He never advanced any claim to the pecuniary rewards derived from either invention. "The only reward I ever expected was the consciousness of advancing science, the pleasure of discovering new truths, and the scientific reputation to which these labors would entitle me."

Some men achieve greatness in a chosen field by cutting off all expenditure of energy in other directions. Henry did not favor this. He will be best remembered by his contributions to electrical science, but his other activities cannot be ignored. Few writers who have treated with the correlation of forces so much as mention Henry's name, but careful study of his work in this field shows him to have been among the very first to form clear ideas of the correlation of forces which extended beyond a mere transformation of one force into another, as magnetism into electricity, or electricity into light. He was looking beyond this into a realm of nature where there was a unitarian force, the power or ability to do work, manifesting itself in many ways. He had a dim vision of energy and came near to grasping it. Had he received any encouragement to pursue these ideas he might have ranked among the more prominent pioneers of the movement, but his neglect to make public his ideas as they occurred to him gave impression of a timid and reluctant acknowledgement of facts, rather than of bold leadership.

Henry's influence upon industry was much greater than its influence upon his work. The brilliance of his contributions to the science of electricity, soon to be applied to the electric telegraph and the motor, swept him into a position of acknowledged leadership in American science. This had an influence upon the introduction of a more intensified study of physics generally into the universities, and this eventually became a benefit to industry.

His work as an organizer of science, although difficult to evaluate, is perhaps as important as his scientific discoveries. He was severely practical in his views and objectives. The advance of science demands that observation and experiment and theoretical discussion should progress in parallel lines. Without organization one of these teams on whose joint exertions progress depends may outrun the others. Henry had observed the retardation of the steam engine through the want of theoretical knowledge and metallurgical advances. He lived to see his own experimental work greatly extended to practical applications. In addition, his work in organization stimulated science by distribution of publications and encouraged scientists by professional recognition of good work. But he made no attempt to affiliate science with the public consciousness; he omitted to make provision for coordinating science with other human activities. By failing to provide for public recognition of the social and political implications of science, he unconsciously fostered a withdrawal of scientists from public affairs, which was bad for both science and society. Henry was conscious of the necessity for the integration of science and industry, but the more general ways in which science was useful to society were not yet appreciated.

Those who take the view that society was already becoming disrupted by forces indiscriminately released by science within it should find much to admire in Henry's efforts to create methods of organization for those forces, so that rational methods might be found for promoting and utilizing them. Observe, however, the limited scope of his measures. He could not foresee the present state of science in the community, and made no efforts to provide for it.

After his death, one after another, high-minded and intelligent men spoke of him with reverence as a man, and spoke of his genius with conviction. Those who knew him best have left a picture of a man of humble loves and great goodwill, warm in charity with the world, who had desired to serve science unselfishly. When placed on trial for his character as a man of honor and a man of science, he defended himself in these simple words:

> My life has been principally devoted to science and my investigations in different branches of physics have given me some reputation in the line of original discovery. I have sought, however, no patent for invention and solicited no remuneration for my labors, but have freely given the results to the world; expecting only in return to enjoy the consciousness of having added by my investigations to the sum of human knowledge. The only reward I ever expected was the consciousness of advancing science, the pleasure of discovering new truths, and the scientific reputation to which these labors would entitle me.

This modest expectation was indeed his chief reward. His countrymen came slowly to recognize his merit. Abroad, recognition came only after his death when the Smithsonian Institution reprinted and distributed the record of his scientific work. With the authority of the great Institution to recommend them, the volumes received a more respectful treatment than the stray copies of Silliman's *Journal* or the annual volumes of the Philosophical Society's *Transactions* had commanded. Since that time there has been, in Britain especially, an appreciation of Henry's work, as several quoted passages in this book testify.

Henry's modest ambitions were mainly fulfilled, and he lived a contented life. His personal character was marked by rectitude, simplicity, straightforwardness, and perseverance, all tempered by warm human understanding. As a man he was valued by his friends, but his fame rests upon his great scientific accomplishment, which stands forever in the record for all to read.

APPENDIX

The following tables were compiled from Joseph Henry's paper "On the Application of the Principle of the Galvanic Multiplier to Electromagnetic Apparatus, and also to the Development of Great Magnetic Power in Soft Iron, with a Small Galvanic Element," *Silliman's American Journal of Science*, January 1831, Vol. XIX, pp. 400–408; also in *Scientific Writings of Joseph Henry*, Vol. I, pp. 37–49.

These tables are intended to show in clear and organized form exactly what Henry did in these experiments, to allow the reader to see at a glance the data with which Henry was working, and to get a greater insight into Henry's approach to his work. A modern physicist expresses his data in tables or graphs almost automatically; Henry did not do this, and the figures in these tables are assembled from the paragraphs of his text. Values not given by Henry in his paper but calculated from his description are enclosed in parentheses. Uncertain values are indicated by a question mark. The density of iron was taken as 4.5 oz./cu. in. These tables were compiled by Herbert S. Bailey, Jr.

APPENDIX

Table 1. Experiments of Group I.

Expt.	Battery used (Cu and Zn in dilute acid)	Length of windings on magnet	Number of windings	Size of iron horseshoe	Weight of magnet	Max. wt. lifted	Max. wt. lifted / wt. of magnet
(A)	"small galvanic plates in dilute acid"	35 ft. 400 turns	1	¼ in. diameter 9 in. long (?)	(2 oz.)?	more powerful than conventional magnet of the same size	?
(B)	2½ sq. in. of Zn	30 ft.	1	½ in. diameter 10 in. long	(8.8 oz.)	14 lbs. (224 oz.)	(25)
(C)	same as (B)	30 ft.	1	same as (B)	(8.8 oz.)	39 lbs. (624 oz.)	"over 50" (71)
(D)	same as (B)	two 30-ft. windings	2	same as (B)	(8.8 oz.)	28 lbs. (448 oz.)	(51)

Table 2. Experiments in Group II. The magnet was "a small soft iron horseshoe, ¼ of an inch in diameter, and wound with about 8 feet of copper wire."

Expt.	Size of battery electrodes	Length of connection to magnet	Approx. resistance of connection (ohms)	Maximum weight lifted
3	2" x 2" of Cu 2" x 2" of Zn	1060 ft.	(5)	"scarcely observable"
4(a)	4" x 7" of Zn surrounded by Cu	0	(0)	4½ lbs. (72 oz.)
4(b)	same as 4(a)	1060 ft.	(5)	about ½ oz.
5	same as 4(a)	550 ft.	(2.5)	2 oz.
6	same as 4(a)	550 ft. two wires in parallel	(1.25)	4 oz.
7(a)	25 double-plates in a Cruickshank's trough. Same area of Zn as in 4(a).	0	(0)	7 oz.
7(b)	same as 7(a)	1060 ft.	(5)	8 oz.

APPENDIX

TABLE 3. Experiments in Group III. For reference the coils are numbered from A to I from one end of the magnet to the other. The magnet weighed 21 lbs., the armature 7 lbs. Note that if Henry had plotted the values of maximum weight lifted against number of coils used, he would have obtained a typical magnetic saturation curve. He was content to say, "This is probably the maximum of magnetic power which can be developed in this horse-shoe," and he did not investigate this phenomenon further.

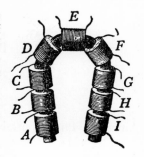

Expt.	Battery size	No. of coils connected in parallel	Coils used	Max. wt. lifted (lbs.)	Max. wt. lifted / wt. of magnet
8		1	Each separately in turn	7	(0.33)
9	Two concentric copper cylinders with a zinc cylinder between them, exposing ⅖ sq. ft. of zinc.	2	D, F	145	(6.9)
10		2	A, I	200	(9.5)
11		3	A, E, I	300	(14.3)
12		4	A, B, H, I	500	(23.9)
13		6	A, B, C, G, H, I	570	(27.2)
14		9	all	650	(31.0)
15(a)	Zn plate 12" x 6" surrounded by Cu	9	all	750	(35.8)
15(b)	28 plates of Zn and Cu, each 8 inches square	9	all	750	(35.8)

BIBLIOGRAPHY

American Philosophical Society. *Transactions*. Vols. 6–8. Philadelphia, 1834–1842.
Ames, J. S. Certain aspects of Henry's Experiments on Electromagnetic Induction. *Science*. Vol. 75, 1932, p. 89.
Bates, R. S. *Scientific Societies in the United States*. New York, 1945.
Burlinghame, Roger. Magnetic Communication. *March of the Iron Men*. New York, 1943, p. 261.
Crowther, J. G. *Famous American Men of Science*. New York, 1937.
Dall, W. H. *Spencer Fullerton Baird: a Biography*. Philadelphia, 1915.
Davenport, W. R. *Biography of Thomas Davenport: the Brandon Blacksmith*. Montpelier, 1929.
Dickerson, E. N. *Joseph Henry and the Magnetic Telegraph: an address*. Princeton, 1885.
Fahie, J. J. *History of the Electric Telegraph to the Year 1837*. London, 1884.
Fahie, J. J. Magnetism, Electricity, and Electromagnetism up to the crowning work of Michael Faraday in 1831. *Journal of the Institution of Electrical Engineers*. Vol. 69. London, November 1931.
Faraday, Michael. *Experimental Researches in Electricity*. 3 vols., London, 1839.
Fulton, J. F., and E. H. Thomson. *Benjamin Silliman, 1779–1864: Pathfinder in American Science*. New York, 1948.
Gherardi, B. Henry as an Electrical Engineer. *Bell System Technical Journal*. Vol. 11, 1932, p. 327.
Goode, G. B. The Origin of the National Scientific and Educational Institutions of the United States. *Annual Report of the American Historical Society* for 1889. New York, 1890.
Goode, G. B. The Secretaries. *Smithsonian Institution:1846–1896*. Washington, 1897.
Gray, Asa. *Letters*. Edited by J. N. Gray. 2 vols., Boston, 1893.
Harlow, Alvin. *Old Wires and New Waves*. New York, 1936.
Haydon, F. S. *Aeronautics in the Union and Confederate Armies*. Vol. 1, Baltimore, 1941.
Henry, Joseph. *Scientific Writings*. (Abbr. *S.W.*) 2 vols., Smithsonian Institution, Washington, 1886.

BIBLIOGRAPHY

Henry, Mary A. America's part in the Discovery of Magneto-Electricity. *Electrical Engineer.* Vol. 13, New York, 1892.

Henry, Mary A. The Electro-Magnet, *Electrical Engineer.* Vol. 17, New York, 1894.

Jaffe, Bernard. Joseph Henry (1797–1878): the United States establishes a new incubator for science. *Men of Science in America,* 1944, New York, p. 179.

Jones, Bence. *Life and Letters of Faraday.* 2 Vols., Philadelphia, 1870.

Kennelly, A. E. Work of Joseph Henry in Relation to Applied Science and Engineering. *Science.* Vol. 76, 1932, p. 1.

Magie, W. F. Joseph Henry. *Review of Modern Physics.* Vol. 3, October 1931, p. 465.

Magie, W. F. Joseph Henry: Pioneer in Space Communication. *Proceedings of the Institute of Radio Engineers.* Vol. 30, No. 6, June 1942.

McAlister, A. M. *Amos Eaton: Scientist and Educator.* Philadelphia, 1941.

Mendenhall, T. C. The Henry. *Atlantic Monthly.* Vol. 73, May 1894.

Mitchell, S. W. Address. *Semi-Centennial Anniversary of the National Academy of Sciences,* Washington, 1913, p. 89.

Morse, Samuel F. B. *Letters and Journals,* edited by F. L. Morse. 2 vols., Boston, 1914.

Morse, Samuel F. B. *Memorial.* Printed by Act of Congress, Washington, 1875.

Mottelay, P. F. *Bibliographical History of Electricity and Magnetism.* London, 1922.

Newcomb, S. *Reminiscences of an Astronomer.* Boston, 1913.

Odgers, M. M. *Alexander Dallas Bache: Scientist and Educator.* Philadelphia, 1947.

Osborne, H. S., and A. M. Dowling. The Electrical Discoveries of Joseph Henry. Supplement to the *Bell System Technical Journal,* July 1932.

Pope, F. L. Life and Work of Joseph Henry. *Journal of the American Electrical Society.* Vol. 2, 1878, p. 134.

Prime, S. I. *Life of Samuel Morse.* New York, 1875.

Rodgers, A. D. *John Torrey: a Study of North American Botany.* Princeton, 1942.

Royal Society. *Catalogue of Scientific Papers, 1800–1863.*

Smith, E. F. *Life of Robert Hare: an American Chemist (1781–1858).* Philadelphia, 1917.

Smithsonian Institution. *Annual Reports, 1846–1879.* Washington.

Smithsonian Miscellaneous Collections, No. 356. *Memorial of Joseph Henry.* Washington, 1881.

Taylor, W. B. *Joseph Henry's Contributions to the Electric Telegraph.* Washington, 1879.

BIBLIOGRAPHY

True, F. W., editor. *History of the First Half-Century of the National Academy of Sciences, 1863–1913.* Washington, 1913.

Wheeler, J. A., and H. S. Bailey. Joseph Henry (1797–1878): Architect of Organized Science. *American Scientist.* Vol. 34, October 1946, p. 619.

INDEX

acoustics of public buildings, 249
adiabatic expansion, 21
Adams, John, 270
Adams, John Quincy, 113, 175
aeronautics in Civil War, 264 ff.
Agassiz, Louis, 199, 277, 295, 323
Agricultural Bureau, 209
Agriculture, Dept. of, 320
Airey, Sir Robert, 261
Alaskan exploration, 200
Albany, 11, 19, 123
Albany Academy, 16, 19, 24, 28, 43, 96
Albany Institute, 20; paper on Rarefaction of Air, 22; on Topography of New York, 27; on Flame, 40; on Aurora Borealis, 73
Albany Lyceum, 20
Alcoholic spirit rectification, 302
Alexander, Alexander, 43
Alexander, Alexander S., 28
Alexander, Harriet, see Henry, Harriet A.
Alexander, Hugh, 6
Alexander, Stephen, 7, 43, 72, 75, 104, 166
American Academy of Arts and Sciences, 270
American Association for the Advancement of Science, 274; paper on the Imponderables, 142, 161; on Phosphorescence, 151; on Cohesion, 247; on Acoustics, 249
American Philosophical Society, 106, 168, 270; Henry's papers, 109, 110, 130, 131, 132, 151, 153, 155, 159, 167
Ames, Joseph S., 91, 283
Ampère, Andre, 35, 42
Antwerp, 123
Arago, Dominique F., 37
Arctic expedition, 200
Association of American Geologists, 275
atomic physics, 159
Atwater, Caleb, 191
Aurora Borealis, 73, 151
Avebury, Lord, see Lubbock, John
Ayrton, William E., 92

Bache, Alexander D., influence of education, 100; urges publication, 110; companion in Europe, 112; on Smithsonian Board, 176; invites support for Henry, 179; pleads with Henry, 189; friendship with Henry, 244; suggests national scientific board, 258; appointed to Permanent Commission, 259; aids in founding National Academy of Sciences, 277
Bache, Hartman, 267, 268
Bailey, Herbert S., 341
Bain's chemical telegraph, 224
Baird, Lucy, 188
Baird, Matthew, 250
Baird, Spencer F., 186, 189, 199, 287
ballistics, 152
balloons, 264
Barker, George F., 317
Barlow, Joel, 174
Barlow, Peter, 55, 228
Beck, Lewis, 18, 42, 58
Beck, Romeyn, 17, 19, 24, 30, 148
Becquerel, Antoine C., 150
belief, religious, 296 ff.
Bell, Alexander Graham, 312
Berliner, Emile, 311
Bernese Oberland, 323
Berzelius, John J., 159, 161
Biot, Jeane Baptiste, 122
Blake, William P., 232
Bompas, C. C., 155
Boscovitch, R., 160
Boursel, Charles, 313
Boyle, Robert, 14
Brackett, Professor, 98
Brewster, Sir David, 124, 167, 179
bribery, attempted, 301
British Association for the Advancement of Science, 125, 167, 204
Brunel, Sir Isambard, 116
Brussels, 123
building materials, 247
Bureau of Ethnology, 202
Burr, David H., 27

[347]

INDEX

Cajori, F., 76
Calhoun, John C., 113, 185
Cameron, Henry C., 49, 169, 178, 184
Cameron, Simon, 266
capillarity of metals, 152
Capitol buildings, 113, 247, 249
Carlyle, Thomas, 156
Carpenter, George W., 43, 86
Carpenter, William, 159
Catalogue of Scientific Papers, 204
Centennial Exposition, 315, 320
Chase, Samuel F., 223, 266
Childs, Mrs., 307
Chittenden, E. L., 238
Choate, Rufus, 211
Cleaveland, Parker, 74, 88
Clinton, George, 42
coal fuel, 71
cohesion of particles, 153, 247
coinage, 262
College of New Jersey, *see* Princeton University
Colorado canyons surveyed, 201
compass deviation, 261
Conklin, Judge, 23
conservation of energy, 156
Contributions to Electricity and Magnetism, 109, 110, 129, 131, 138
Cooper Union, 313
Copley Medal, 120
Corcoran Gallery, 288, 320
Cornell, Ezra, 216
correspondents, 305
Cotes, Roger, 321
Coulomb, Charles, 32
Cousins, Victor, 100
Crowther, J. G., 76
Cruickshanks trough, 34, 54
Cuyler, Theodore L., 87

Dahlgren, John A., Admiral, 248, 249
Dal Negro, Salvatore, 70
Dana, James Dwight, 273, 278
Dana, James Freeman, 38
Daniell, John F., 119
Darwinian controversy, 293
Davenport, Thomas, 66, 71, 94, 127
Davis, Charles H., Admiral, 258, 277
Davis, Jefferson, 235
Davy, Sir Humphrey, 19, 40, 149, 300
De La Rive, August A., 41, 122
Descartes, Renè, 157
De Witt, Richard V., 20, 112, 114
De Witt, Simeon, 29
Dickerson, Edward, 99, 108
Dod, Samuel B., 72, 143, 291

Doty, John F., 12
Douglas, Stephen A., 209, 229
Downing, Andrew, 188
Dufay, C. F., 31
Dunlap, William, 93, 173

Eaton, Amos, 18, 26, 27, 127
eddy currents, 135
Edinburgh, 124
Edmundson, Thomas, 71
education, 100 ff., 322
electric eel, 138
electric telegraph, *see* telegraph, electromagnetic
electricity, state of, 30 ff.
electromagnetic relay, 108, 120, 335
electromagnetic shielding, 135, 335
electromagnets, Bowdoin College, 74, 88; Henry's, 42, 48 ff., 56, 61, 105; Moll's, 46; Penfield Iron Works, 66, 75; Princeton, 105; Sturgeon's, 38; Yale, 67, 75
Ellert, Charles, 201
Emerson, Ralph W., 156, 171
energy, conservation of, 156
England, visits to, 114, 321
English thought, 115
Erie Canal, 25, 42
Espey, James P., 195
Ethnology, Bureau of, 202, 321
European visits, 112, 321
Evans, Oliver, 120
exploration of the West, 199

Faraday, Michael, introduced to science, 15; discovers mutual induction, 76 ff.; parallels with Henry, 78, 136; difference in method, 88; self induction, 91, 109; non-inductive wirings, 111; meets Henry, 117; as lecturer, 118; ray vibrations, 141; on screening effects, 136
Felton, Cornelius C., 229
Fillmore, Millard, President, 247
Fleming, J. A., 91
fog horns, 253
Fool of Quality, 10
Foote, Elial T., 20
Forbes, James D., 87
Franklin, Benjamin, 31, 37, 73, 120
Franklin Institute, 106, 110, 120, 153, 168, 273
Fresnel lenses, 251

Gale, Leonard D., 127, 215, 219, 224, 230
Galvani, Luigi, 33
Galway, N.Y., 8
Gautier, Alphonse, 166
Gay-Laussac, Joseph L., 122

[348]

INDEX

Gibbs, Wolcott, 262, 278
Gibson, William J., 142, 337
Gilbert, William, 31
Gould, B. A., 279
Gray, Asa, 99, 159, 205, 291, 294
Gray, Stephen, 31
Green, Jacob, 50, 98, 307
Gregory's *Lectures on Experimental Philosophy*, 14
Guericke, Otto von, 31
Guyot, Arnold, 195, 323

Haldeman, Samuel, 293
Hall, James, 231
Hamilton, Sir William R., 333
Hammond, John D., 24
Hansteed magnetic needles, 72
Hare, Robert, 38, 40, 50, 82, 106, 184, 319
Harris, Sir William S., 114
Harvard University, 282
heat radiation, 250
Helmholtz, H. L. F. von, 140
Hendrie, *see* Henry, William
Henry, Ann Alexander, 6, 7
Henry, Caroline, 46
Henry family, 5
Henry, Harriet Alexander, 44, 113, 124
Henry, Helen, 46
Henry, Joseph, birth, 7; attends schoool, 8; apprenticed to watchmaker, 12; interest in theatre, 12; attends Albany Academy, 16; teaches district school, 16; private tutor, 17; studies medicine, 18; meets Eaton, 18; chemical experiments, 19; gifts to Albany Institute, 20; on steam, 21; on rarefaction of air, 22; studies mathematics, 23; surveys road, 23; health, 24; professor at Albany Academy, 24; attitude toward money, 24, 64, 298; sense of duty, 26; geological survey of New York, 27; meteorology of New York, 29; sees electromagnet, 38; financial obstacles, 39; on Flame, 41; first electromagnet, 41; insulates wire, 42; letters to Harriet, 45, 113, 124; marriage, 45; increases power of electromagnet, 47; first "quantity" magnet, 50; electromagnetic telegraph, 52; approaches Ohm's law, 56; first "intensity" magnet, 56; magnetic saturation of iron, 59; approaches magnetic circuit, 61; magnet at Penfield, 66; reciprocating motor, 69; investigates aurora borealis, 73; magnet for Bowdoin College, 74; contemplates dynamo, 76; parallels with Faraday, 78, 136; experiments in induction, 81; failure to publish results, 85, 159, 165; discovers self-induction, 89; meets William Dunlap, 93; appointed to Princeton, 96; personal appearance, 99, 286; on education, 100 ff.; installs telegraph, 107; deposition in Morse v. O'Reilly, 108; discovers non-inductive wiring, 111 ff.; impressions of Washington, 113; at Plymouth Dockyard, 114; at Stonehenge, 116; meets Faraday, 117; impressions of London, 117; on Faraday, 118; at King's College, 119; recommended for Copley Medal, 120; in Paris, 121; views on French life, 121; on financial depression, 122; escapes injury, 123; visits Brussels and Antwerp, 123; returns to London, 123; inspects railroad, 124; visits Scotland, 124; addresses British Assoc., 125; returns home, 126; on Davenport's motor, 127; meets Morse, 128; resumes study of induction, 130; on lateral discharge from Leyden jar, 131; discovers principle of transformer, 132; treats facial paralysis, 134; investigates eddy currents, 136; discovers currents of high orders, 137 ff.; discovers oscillatory nature of Leyden jar discharge, 139 ff.; theory of light, 141; induction at a distance, 142 ff.; lack of appreciation, 146; on sound, 150; on phosphorescence, 151; measures velocity of projectiles, 152; capillarity of metals, 152; investigates cohesion, 154, 247; investigates molecular forces, 153 ff.; on natural motors, 155 ff.; on transformation of forces, 156 ff.; on atomic physics, 159; on the imponderables, 161 ff.; experiments with lightning, 165; on solar physics, 166; position in 1847, 169; plan for the Smithsonian, 176; offered post at Smithsonian, 179; reluctance to accept, 180; appointed secretary, 179; offered chairs at universities, 184, 185; offered presidency of Princeton, 185; generosity to students, 188; opposes expense on buildings, 188; threatens to resign, 189; plans publications, 190; disapproves of copyright, 192; organizes exchanges, 193; organizes meteorological observation, 194; daily weather map, 196; articles on meteorology, 197; promotes specialists' studies, 199; encourages explorations, 199; interest in ethnology and anthropology, 202; projects catalogue of

INDEX

Henry, Joseph (*continued*),
scientific papers, 203; starts national herbarium, 205; as financial officer, 206; defends institution funds, 209; opposes large library, 210; controversy with Jewett, 212; controversy with Morse, 214 ff.; vindicated by regents, 231; British tribute, 233; friendships with Andrew Johnson and Jefferson Davis, 235; on the Union, 237; opinion of Lincoln, 238; incurs ridicule, 242; accused of spying, 243; correspondence and papers lost, 244; tests building materials, 246; views on cohesion, 247; on acoustics, 249; on radiant energy, 250; appointed to Light-House Board, 251; experiments with oils, 251; experiments with fog horns, 253; differs from Tyndall, 255; appointed chairman of Light-House Board, 255; appointed to Permanent Commission, 259; committees of National Academy, 260; work on aeronautics, 264 ff.; organizes science, 270 ff.; and the A.A.A.S., 274; the National Academy of Sciences, 278; appeals for recognition of scientists, 279; awarded university degrees, 282; character, 285; love of literature, 289; attitude to religion, 291; outlines belief, 296; encounter with spiritualist, 308; on will o' the wisp, 307; encourages Alexander Bell, 312; transfers library and museum, 320; second trip to Europe, 321; receives testimonial fund, 326; final address to National Academy, 326; last illness, 326; contributions to science, 330 ff.
Henry, Mary, 46, 86, 289, 321
Henry Memorial Lectures, 283
Henry Mountains, 201
henry, the, 92
Henry, William, 6
Henry, William, Jr., 6
Henry, William Alexander, 46, 244
Herschel, Sir William, 166
Hertz, Heinrich, 144
Hilgard, E., 264
Holden, Edward J., 276
House's printing telegraph, 224
Hun, Thomas, 121
Huxley, Thomas, 322
Huyghens, Christian, 157
hypotheses, 164, 331

Imponderables, Theory of the so-called, 142, 161

Indians, reports on, 201
induction, mutual, 76 ff.
induction, self, 89 ff.
inland revenue frauds, 263
Institut des Sciences, 122
insulated wire, first used, 42
International Congress of Electricians, 92

Jacobi, M. H., 147
James, Henry, 23, 123
Jeans, Sir James, 142
Jewett, Charles C., 212
Johnson, Andrew, President, 235
Joule, J. Prescott, 71, 164
Junto, 270

Keeley, Robert, 303
Kelvin, Lord, *see* Thomson, William
King's College, London, 119

Lagrange, Joseph L., 23
Laplace, Pierre S., 206
Leyden jars, 31, 131
Library of Congress, 319
light, 141, 145, 151, 160
Light-House Board, 251 ff.
lighthouses in Scotland, 124
lightning, 165
Lincoln, Abraham, President, 238 ff., 242, 244, 308
Lincoln, Mary Todd, 308
Lincoln, N. S., 326
liquids, cohesion of, 154
literature, love of, 289
Lodge, Sir Oliver, 145
London, 116, 117, 123
Loomis, Elias, 73, 195
Lowe, Thaddeus S. C., 265
Lubbock, John, Lord Avebury, 322
Lyell, Charles, 293

Maclean, John, 97
McCullach, Hugh, 260
magnetic intensity at Albany, 72
magnetic saturation of iron, 59
March, Dr., 18
Marconi, Guglielmo, 146
Mascart, E. E. N., 92
Mason, Charles, 225
Mason, James W., 229
materialism in society, 301
mathematics in physical studies, 333
matter, constitution of, 159
Maxwell, Clerk, 141, 144
Mayer, Julius, 157
Meacham, Representative, 213

[350]

INDEX

Meade, George G., General, 245
Meads, Orlando, 19, 97
medical profession and science, 26
Meigs, Montgomery C., General, 249, 282
Melloni, Macedoine, 122
meteorology, 30, 73, 165, 194 ff.
metric system, 260
militarism, 121
Mills, Clark, 329
Mississippi River survey, 201
Mitchell, S. Weir, 129, 326
molecular forces, 153 ff.
Moll, Gerard, 46
Morse, Samuel F. B., 56, 127, 214 ff.
Morse versus O'Reilly, 108, 223
motor, Davenport's rotary, 127; Henry's reciprocating, 69, 127
motors, natural, 155
mutual induction discovered, 76

National Academy of Sciences, 260, 277, 316, 326
National Gallery of Art, 321
National Herbarium, 205
National Institution, 175
National Museum, 320
National Zoological Park, 321
nature, love of, 323
Negro question, 124, 237
New York Avenue Presbyterian Church, 295
New York State roads, 25
New York State, topography of, 27
Newcomb, Simon, 37, 285, 305, 310, 327
Newton, Sir Isaac, 46, 61, 125, 142, 331
Nollet, Abbé J. Anthoine, 31
non-inductive wiring, *see* wiring, non-inductive
Nott, Eliphalet, 16, 182

occultism, 308
Oersted, Hans Christian, 35, 84, 146, 293
Ohio River survey, 201
oil experiments, 251 ff.
O'Reilly's telegraph, 223
Origin of Species, 293
Orton, William, 316
oscillations, electrical, 139
Owen, Robert D., 175

Pacific Railroad surveys, 201
Paris, 121
Pasteur, Louis, 3, 302
Paterson, R. M., 153
Patterson, Joseph, 296
Peake, "Old," 311

Pedro, Emperor of Brazil, 315
Penfield Iron Works, 66, 75
Pennsylvania, University of, 184
Percy, Earl of, 172
Permanent Commission of Navy, 258
Phelps, Israel, 8
phonometer, 253, 311
phosphorescence, 150
Pierce, Benjamin, 262, 279
Pierce, James A., 229
Plateau, Joseph, 155
Plymouth Dockyard, 114
Poinsett, Joel R., 175
Pouillet, Claude S. M., 68, 90
Powell, John W., 201
Priestley, Joseph, 32
Princeton, U.S.S., 153
Princeton University, 96, 97, 169
projectiles, velocity of, 152

radiant energy, 250
Raffinesque, Constantine, 293
Rankine, W. J. M., 71
Reis, Philipp, 313
relay, electromagnetic, 108, 120, 335
religion and science, 292
Rennselaer, Stephen van, General, 17, 20, 127
Renwick, James, architect, 286
Renwick, James, professor, 72, 98
revenue tax on spirits, 262
rewards of science, 299
Rhone Glacier, 323
Ritchie, William, 53
Rogers, Fairman, 184
Roget, Peter M., 76
Romagnosi, Gian D., 34
Rostrum, The, 13, 15
Royal Commission on science teaching, 322
Royal Institution, 78, 117, 118
Royal Society, 117, 120, 204
Rumford, Count, *see* Thompson, Benjamin
Rush, Richard, 117

St. Elmo's fire, 126
St. Paul's Cathedral, 117
Saturday Club, Washington, 282
Savary, Felix, 139
Saxton, Joseph, 245, 282
Schweigger, Johann, 27, 47, 53
science and religion, 292
science and technology, 299
science, Henry's contributions to, 330 ff.
science in the nineteenth century, 271

[351]

INDEX

science, organization of, 270, 338
scientific training, difficulties of, 17
scientific societies, 270
scientists, appeal for recognition of, 279
scientists, place in society, 299, 318
Scott, Sir Walter, 125
Secchi, P. A., 167
self-induction, 89, 109
Shaffner, Taliafero P., 225
Shephard, Charles U., 68
shielding, elecric, 335
Shubrick, William B., Admiral, 255
Silliman, Benjamin, 67, 75, 98, 115, 191
Silliman, Benjamin, Jr., 282
Smithson, Hugh, 117, 172 ff., 244
"Smithsonian" balloon, 264
Smithsonian Contributions to Knowledge, 191
Smithsonian Institution, Henry's plan, 177; qualifications for Secretary, 178; standing among scientific institutions, 186; building, 187; finances, 206; library controversy, 212; staff provides disinfectants, 241; fire at, 244; staff work in Civil War, 257; aides to other organizations, 273; tower used for experiments, 303; correspondents, 305; Newcomb's first visit, 310; departments transferred, 319
Smithsonian Miscellaneous Collections, 191
Socrates, 328
solenoid introduced, 42
Solomons, S. S., 311
sound waves, 253
South Carolina University, 282
specific gravity of alcoholic spirits, 262
spirit rectification, 302
spiritualism, 307
Staten Island Laboratory, 325
steam, mechanical and chemical effects, 21
steamboat controversy, 125
Steinheil, Karl A., 223
Stokes, Sir G. G., 151, 253, 322
Stonehenge, 116
Story, William W., 329
Sturgeon, William, 38, 39, 67
sunspots, 166
surveys of United States, 199
Swammerdam, J., 33
swindles, exposure of, 302

Taney, Robert Brooke, Chief Justice, 229
Taylor, William B., 282
technology and science, 299

telegraph, electromagnetic, 55, 63, 107, 128, 214 ff., 335
telephone, 312, 316, 336
Ten Eyck, Philip, 53, 61, 75, 147
thermal telescope, 167
Thomson, Benjamin, Count Rumford, 250
Thomson, Sir J. J., 146
Thomson, William, Lord Kelvin, 141, 144, 262, 300, 315, 324
Topographical Sketch of New York, 27
Torrey, John, 17, 98, 103, 179, 205, 237, 244, 277
transformer, electrical, 132, 335
Trinity House, 324
Trowbridge, W. P., 215
Tully, Dr., 18
Tyler, G., 326
Tyler, John, President, 153
Tyndall, John, 254, 255, 324

Union College, 43
U.S. Corps of Civil Engineers, 23
University of New York Regents, 30

Vail, Alfred, 216, 217
Valéry, Paul, 284
Valtmare, Alexandre, 193
Van Buren, Martin, President, 113
Van Wyck, P. C., 150
Vaughan, Henry, 117
Vethake, Henry, 96
Volta, Alessandro, 33

Walker, Sears C., 218
Ward, Samuel, 309
Warren, J. C., 274
Washington, 113, 174, 186, 236
Washington Philosophical Society, 255, 282, 296, 315
Weather Bureau, 194 ff.
Webster, J. W., 82
Welles, Gideon, 259
Welling, James C., 93, 301
Western Union Telegraph Company, 200, 316
Westminster Abbey, 118
Wheatstone, Sir Charles, 108, 119, 120, 223
Whitney, Eli, 116
will o' the wisp, 307
Wilson, Henry, 277
Winthrop, John, 270
wiring, non-inductive, 111
Wise, John, 197, 264, 266
Wordsworth, William, 289

Yale University, 67, 75